Chemistry, Meteorology and the Function of Digestion Considered with Reference to Natural Theology

WILLIAM PROUT

CAMBRIDGE
UNIVERSITY PRESS

CAMBRIDGE UNIVERSITY PRESS

Cambridge, New York, Melbourne, Madrid, Cape Town, Singapore,
São Paolo, Delhi, Dubai, Tokyo

Published in the United States of America by Cambridge University Press, New York

www.cambridge.org
Information on this title: www.cambridge.org/9781108000666

© in this compilation Cambridge University Press 2009

This edition first published 1834
This digitally printed version 2009

ISBN 978-1-108-00066-6 Paperback

THE BRIDGEWATER TREATISES

ON THE POWER WISDOM AND GOODNESS OF GOD

AS MANIFESTED IN THE CREATION

TREATISE VIII

CHEMISTRY METEOROLOGY AND THE FUNCTION

OF DIGESTION

BY WILLIAM PROUT M.D. F.R.S.

[SECOND EDITION]

Μὴ ὑπαρχούσας γὰρ ἁρμονίας, καὶ ΈΠΟΨΙΟΣ ΘΕΙΑΣ
περὶ τὸν κόσμον, οὐκ ἂν ἐδύνατο ϛυνεῖμεν ἔτι καλῶς ἔχοντα τὰ
ἐγκεκοσμαμένα.   HIPPODAMUS DE FELICITATI.

CHEMISTRY METEOROLOGY

AND THE FUNCTION OF DIGESTION

CONSIDERED WITH REFERENCE TO

NATURAL THEOLOGY

BY

WILLIAM PROUT M.D. F.R.S.

FELLOW OF THE ROYAL COLLEGE OF PHYSICIANS

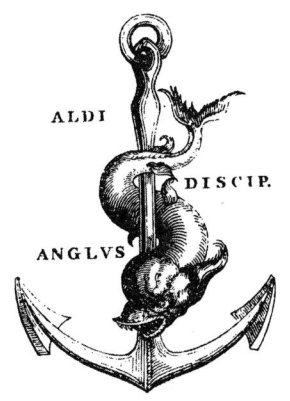

ALDI DISCIP. ANGLVS

LONDON

WILLIAM PICKERING

1834

DAVIES GILBERT ESQ.

LATE PRESIDENT OF THE ROYAL SOCIETY

This Volume

IS

RESPECTFULLY INSCRIBED

NOTICE.

The series of Treatises, of which the present is one, is published under the following circumstances:

The Right Honourable and Reverend Francis Henry, Earl of Bridgewater, died in the month of February, 1829; and by his last Will and Testament, bearing date the 25th of February, 1825, he directed certain Trustees therein named to invest in the public funds the sum of Eight thousand pounds sterling; this sum, with the accruing dividends thereon, to be held at the disposal of the President, for the time being, of the Royal Society of London, to be paid to the person or persons nominated by him. The Testator further directed, that the person or persons selected by the said President should be appointed to write, print, and publish one thousand copies of a work *On the Power, Wisdom, and Goodness of God, as manifested in the Creation; illustrating such work by all reasonable arguments, as for instance the variety and formation of God's creatures in the animal, vegetable, and mineral kingdoms; the effect of digestion, and thereby of conversion; the construction of the hand of man, and an infinite variety of other arguments; as also by discoveries ancient and modern, in arts, sciences, and the whole extent of literature.* He desired, moreover, that the profits arising from the sale of the works so published should be paid to the authors of the works.

The late President of the Royal Society, Davies Gilbert, Esq. requested the assistance of his Grace the Archbishop of Canterbury and of the Bishop of London, in determining upon the best mode of carrying into effect the intentions of the Testator. Acting with their advice, and with the concurrence of a nobleman immediately connected with the deceased, Mr. Davies Gilbert appointed the following eight gentlemen to write separate Treatises on the different branches of the subject as here stated:

THE REV. THOMAS CHALMERS, D.D.

PROFESSOR OF DIVINITY IN THE UNIVERSITY OF EDINBURGH.

ON THE POWER, WISDOM, AND GOODNESS OF GOD
AS MANIFESTED IN THE ADAPTATION
OF EXTERNAL NATURE TO THE MORAL AND
INTELLECTUAL CONSTITUTION OF MAN.

JOHN KIDD, M.D. F.R.S.

REGIUS PROFESSOR OF MEDICINE IN THE UNIVERSITY OF OXFORD.

ON THE ADAPTATION OF EXTERNAL NATURE TO THE
PHYSICAL CONDITION OF MAN.

THE REV. WILLIAM WHEWELL, M.A. F.R.S.

FELLOW OF TRINITY COLLEGE, CAMBRIDGE.

ASTRONOMY AND GENERAL PHYSICS CONSIDERED WITH
REFERENCE TO NATURAL THEOLOGY.

SIR CHARLES BELL, K.G.H. F.R.S. L.&E.

THE HAND: ITS MECHANISM AND VITAL ENDOWMENTS
AS EVINCING DESIGN.

PETER MARK ROGET, M.D.

FELLOW OF AND SECRETARY TO THE ROYAL SOCIETY.

ON ANIMAL AND VEGETABLE PHYSIOLOGY.

THE REV. WILLIAM BUCKLAND, D. D. F. R. S.

CANON OF CHRIST CHURCH, AND PROFESSOR OF GEOLOGY IN THE
UNIVERSITY OF OXFORD.

ON GEOLOGY AND MINERALOGY.

THE REV. WILLIAM KIRBY, M. A. F. R. S.

ON THE HISTORY, HABITS, AND INSTINCTS OF ANIMALS.

WILLIAM PROUT, M. D. F. R. S.

CHEMISTRY, METEOROLOGY, AND THE FUNCTION OF DIGESTION, CONSIDERED WITH REFERENCE TO NATURAL THEOLOGY.

His Royal Highness the Duke of Sussex, President of the Royal Society, having desired that no unnecessary delay should take place in the publication of the above mentioned treatises, they will appear at short intervals, as they are ready for publication.

CHEMISTRY has not hitherto been considered
in detail with reference to Natural Theology;
the difficulties, therefore, incidental to a first
attempt, added to those arising from the
nature of Chemistry itself as a science, must
be the apology of the author for numerous
imperfections in this treatise.

The peculiar chemical opinions advanced,
would never have appeared in their present
form; had not the author been strongly impressed with the belief that they are calculated, sooner or later, to bring chemical
action under the dominion of the laws of
quantity; and had he not despaired, under
his professional engagements, of being himself able to submit them to experimental

proof. These opinions, however, have been always introduced as mere illustrations.

The argument of design is necessarily cumulative ; that is to say, is made up of many similar arguments. To avoid repetitions therefore, the illustration of principles rather than of details, has been studied ; and the application of particular facts to the argument, has been often left to the Reader.

London,
February 3, 1834.

TO THE READER.

[SECOND EDITION.]

THE Author in certain parts of this treatise having been misunderstood, is anxious to state more prominently than in the first edition, *that no argument illustrative of design has been founded on the supposed molecular arrangements which he has given;* and that the reality of design in the phenomena of chemistry is no more affected by the truth or falsehood of his theory, than the moral of a fable is affected by the truth or falsehood of its imaginary incidents.

The phenomena of chemistry can neither be represented by figures, nor adequately described to the inexperienced by words. The consideration of chemistry, therefore, with reference to the argument of design, presents peculiar difficulties. After much reflection the author has omitted details which every treatise on the subject will furnish, and has endeavoured to convey some notion of the wonders of molecular action. As the best suited to his purpose, a sketch of his own views is given : and if the uninformed reader learn from this sketch, that the *invisible* operations of chemistry, are at least as wonderful as the *visible* operations of mechanism, the author will attain one of his objects. He will attain another object, if what has been stated, in any way contribute to the advancement of knowledge.

In this edition the introductory observations have been enlarged ; a few errors have been rectified ; and the arrangement of some parts altered : but with the exception of these changes, the work remains essentially the same as in the first edition.

London,
June 7, 1834.

CONTENTS.

INTRODUCTION.

BOOK I.

OF CHEMISTRY.

P. b

BOOK II.

OF METEOROLOGY:

COMPREHENDING A GENERAL SKETCH OF THE CONSTI-
TUTION OF THE GLOBE; AND OF THE DISTRIBUTION
AND MUTUAL INFLUENCE OF THE AGENTS AND ELE-
MENTS OF CHEMISTRY IN THE ECONOMY OF NATURE.

CHAPTER VI.—OF THE ADAPTATION OF ORGANIZED
BEINGS TO CLIMATE; COMPREHENDING A GENERAL
SKETCH OF THE DISTRIBUTION OF PLANTS AND

BOOK III.

OF THE CHEMISTRY OF ORGANIZATION:

DIRECTION TO THE BINDER.

The Binder is desired to place the Map opposite to page 568, in the Appendix.

INTRODUCTION.

OF THE LEADING ARGUMENT OF NATURAL THEO- LOGY ; THAT DESIGN, OR THE ADAPTATION OF MEANS TO AN END, EXISTS IN NATURE.

A FULL exposition of the argument of design does not belong to this Treatise. In these introductory observations, we shall, therefore, confine ourselves to a statement of the argument, as deducible from a simple instance of the adaptation of means to an end, among the objects of nature : we shall then enquire into the validity of the argument of design ; and shall show the conclusions to which that argument leads.

The instance of adaptation of means to an end which we select among the objects of nature ; and the argument which may be deduced from that instance of adaptation, are the following :

P.   B

Animals in cold climates have been provided with a covering of fur. Men in such climates cover themselves with that fur. In both cases, whatever may have been the end, or intention; no one can deny that the *effect*, at least, is precisely the same: the animal and the man are alike protected from the cold. Now, since the animal did not clothe itself, but must have been clothed by another; it follows, that whoever clothed the animal, apparently knew what the man knows, and reasoned like the man; that is to say, the clother of the animal knew that the climate in which the animal is placed, is a cold climate; and that a covering of fur, is one of the best means of warding off the cold: he therefore clothed his creature in this very appropriate material.

The man who clothes himself in fur to keep off the cold, performs an act directed to a certain end; in short, an act of *design*. So, whoever, directly or indirectly, caused the animal to be clothed with fur, to keep off the cold, must likewise have performed an act of *design*.

But, under the circumstances, the clother of the animal, must be admitted to have been also the *Creator* of the animal; and, by extending the argument, the Creator of man himself—of the universe. Moreover, the intelligence the Creator has displayed in clothing the animal,

He has deigned to impart to man; who is thus enabled to recognise his Creator's design.

Such is an instance of those varied adaptations of means to an end, which we behold in the world; and such the train of reasoning, which, to common understandings, appears to show, that these adaptations are the effect of design. There are, however, some men, whose minds are so obtuse, or so singularly constituted, that they maintain all these appearances of design to be unreal; and a brief examination of the pretexts which they have urged for their incredulity, may not be deemed irrelevant.

The opposers of the argument of design may be divided into two classes: those who, denying a First Cause, affect to believe, that all the beautiful adaptations and arrangements we witness in creation, arise from what they term, "the necessary and eternal laws of nature;" and who, in fact, are Atheists, or rather Pantheists, " to whom the laws of nature are as gods:" and those who, without denying the existence of a First Cause, contend, that the adaptations among the objects of nature, cannot be *proved* to be the effect of design; that these objects appear to us well adapted to each other, because we have nothing, besides our own intellects, with which we can compare them; and that the limited powers of the human mind, are a standard altogether inapplicable to the Deity.

The opposers of the argument of design, who assert the existence of " necessary and eternal laws of nature," need no other refutation, than the facts detailed throughout these Treatises. Those who allege that design cannot be *proved*, may be thus answered :

We have been gifted with mental faculties, by which we recognise certain abstract truths, or necessary existences; which abstract truths, or necessary existences, we cannot doubt, without doubting the existence of ourselves. We have been gifted with other faculties, by which we recognise the existence, and compare the properties, of things *external* to ourselves : but, so far as we can discover, neither the existence of these external things, or the existence of their properties, is *necessary;* they might have existed quite otherwise than they are,—they even might never have had any existence. Now our knowledge of the manifestations of design, among the objects of nature, is derived entirely from those faculties, by which we recognise the existence, and compare the properties, of objects external, to ourselves. Hence nothing can be proved, either for, or against, what we term design in nature, by any argument founded on *necessity;* that is to say, by any *à priori* reasoning, founded on mere abstract truths, or necessary existences; and all such attempts must be not only unsatisfactory, but absurd.

At the commencement of this Introduction, we have given a simple statement of the train of reasoning, on which we found our belief of the agency of an intelligent Creator.

An act, performed by ourselves, when directed to a certain end, we term an act of design. Among the objects of nature, we see the *same* end, attained by the employment of the *same* means, we ourselves employ. We are conscious of the will and the power which are requisite for the accomplishment of our own act; and are satisfied regarding the impossibility of that act, without our own, or similar agency. We thence infer, that without some external agency, (implying a will and a power, similar to the will and the power exerted by ourselves), an act, similar to our own act, could not have been accomplished. Our belief then, in the agency of an intelligent Creator, is founded,—

On our recognition of the identity of effects produced in external nature, with effects produced by ourselves; from which identity of effect, we immediately infer identity of purpose, —the *existence of design*, without reference to a designer :

On our consciousness that the purpose effected by us proceeded from ourselves, the designers ; whence we conclude, that the design manifested in external nature must have had a like origin,—

that the manifestation of design, is demonstrative of the *existence of a designer :*

On the pervading character of the design shown among the objects of nature; in which design, man recognises the *creation* of the objects designed; and is thus led to infer the existence of a *Creator.* Now the faculty of reason, which enables man to recognise the Creator of the objects around him; enables him to recognise in that Creator, the *Creator of himself, and of his faculties.* In reasoning, therefore, from his own acts, to those of the Creator of the Universe, though conscious that he is reasoning from the finite to the Infinite; from weakness to Almighty Power;—yet, when he reflects, from whom he has derived his faculty of reason, man feels assured that his own reasoning, when it coincides with the reasoning evinced by his Creator, *can be no other than the same.* Nor founded, as that assurance is, on the constitution of the human mind, can such assurance be impugned; without impugning Him, by whom the human mind has been so constituted.

Thus the argument of design, though not based on necessity, in the strict sense of the term, is of *a validity equal to that of our knowledge of the existence of, and of our connexion with, an external world.* Speculative men may

deny the existence of all things external to themselves ; may even deny their own existence ; but
while they continue to act like other men, it is
not easy to imagine them sincere. We at least,
discard all such speculations, as worthless fal-
lacies, and contend for the *common-sense* view of
the existence and origin of things ;—that design
is design, whether exemplified in the works of
man, or in those of his Maker ; a view which
has been adopted by the wise and good in all
ages ; which has all the probabilities on its side ;
and which alone, of all others, points out to man
his true and natural position, among created
beings. When man, indeed, compares himself
with the universe, his own insignificance appears
quite overwhelming ; but the argument of design
assures him that, insignificant as he is, while he
investigates and approves of the order and harmony around him, he is exerting faculties truly
god-like—that his reason, though limited in degree, must be immortal in kind, and thus differ
from that of the great Architect of all, only in
not being infinite. And hence the proud rela-
tionship in which man justly considers himself
to stand with respect to his Maker ! hence the
grand source of that longing after a future state,
where his knowledge will be consummated, and
where he will no longer " see through a glass
darkly"—notions at once the result and reward

of his reason, and which raise him far above all other animals.

We have endeavoured to illustrate the argument of design, by one of those obvious examples of the adaptation of means to an end among the objects of nature, which impress on man a belief in the existence of design, and of a Designer. Compared, however, with the extent of creation, the instances, numerous as they appear, in which man is thus able to trace the designs of his Creator, are really few. Man not only sees means directed to certain ends, but ends accomplished by means, which he is totally unable to understand. He also sees, every where, things, the nature, and the end, of which, are utterly beyond his comprehension; and respecting which, he is obliged to content himself, with simply inferring the existence of design.

The argument of design, therefore, in its general sense, embraces at least three classes of objects :—

1. Those objects, regarding which, the reasoning of man coincides with the reasoning evinced by his Creator; as in the simple adaptation of clothing above mentioned: or those objects, in which, man is able to trace, to a certain extent, his Creator's designs; as in various phenomena amenable to the laws of quantity; viz. mechanics, &c.

2. Those objects, in which, man sees no more than the preliminaries and the results, or the end and design accomplished; without being able to trace, through their details, the means of that accomplishment; as in all the phenomena and operations of chemistry.

3. Those objects, in which, design is inferred, but in which the design, as well as the means by which it is accomplished, are alike concealed; as in the existence of fixed stars; of comets; of organic life: and indeed in all the great and more recondite phenomena of nature.

The intention of these Treatises, is to point out the various evidences of design, among the objects of creation; and to deduce from them, the existence, and the attributes of the Creator. The following pages are occupied, more particularly, with the illustration of the evidences of design, in objects belonging to the second, of the three classes, above mentioned; with those, namely, in which design is apparent, though we cannot trace the means by which that design is accomplished.

BOOK I.

OF CHEMISTRY.

PRELIMINARY OBSERVATIONS ON THE RANK OF CHE-
MISTRY AS A SCIENCE; AND ON ITS APPLICATION
TO THE ARGUMENT OF DESIGN.

" CHEMISTRY does not afford the same species of
argument (in favour of design) that mechanism
affords, and yet may afford an argument in a
high degree satisfactory."* This remark of the
excellent Paley, has been made by him with
reference only to a particular subject; but the
following sketch, pointing out the grounds upon
which chemistry as a science is founded, and
the rank which it holds among the departments of human knowledge, will at the same
time, show the general truth of the remark.

An elaborate enquiry into the origin and
nature of human knowledge, would be quite

---

* Natural Theology, chap. vii.

misplaced here. We shall content ourselves, with simply considering it of the two kinds, described in the introduction, viz. : a knowledge of what *must be;* that is to say, of what we cannot conceive either not to exist, or to exist otherwise than as it is ; and which is therefore founded upon reason (or *necessity*) : and a knowledge of what simply *is*, but how or why we know not ;. and for the existence of which, therefore, we have no authority beyond our own consciousness, or the evidence of our senses.

Of these, the only instance of the first kind which particularly concerns us at present, is the knowledge of *quantity*, and its relations in general : of the second, that of certain natural phenomena ; the consideration of which, constitutes the proper subject of the present volume.

The fundamental differences between these two great branches of human knowledge, as well as their consequences, cannot perhaps be more strikingly illustrated, than in the following familiar comparison by a celebrated writer. "A clever man," says Sir J. Herschel, "shut up alone, and allowed unlimited time, might reason out for himself all the truths of mathematics, by proceeding from those simple notions of space and number, of which he cannot divest himself without ceasing to think : but he would never tell by any effort of reasoning, what would become of a lump of sugar, if immersed in water ;

or what impression would be produced on his eye, by mixing the colours yellow and blue ;"* results which can be learnt only from experience.

Thus then, the extremes of human knowledge may be considered as founded on the one hand, purely upon reason ; and on the other, purely upon sense. Now, a very large portion of our knowledge, and what in fact may be considered as the most important part of it, lies between these two extremes, and results from a union or mixture of them ; that is to say, consists of the application of rational principles, to the phenomena presented by the objects of nature.

With respect to knowledge founded upon *reason*, we are so constituted, that whether we contemplate, in the abstract, those primary notions of space, time, force, &c. above alluded to ; or whether we view them in connexion with the objects of sense around us, we cannot divest them of *quantity*, which seems to be involved in their very essence. Quantity and its relations, therefore, in some shape or other, enter as a necessary element, into by far the greater portion of human knowledge. Now the primary relations of quantity are exceedingly simple ; one quantity may be equal to another; or it may be greater, or less ; but we can conceive no other relation. Hence all the operations of the mathe-

* Discourse on the Study of Natural Philosophy, p. 76.

matics—the science of quantity and its relations —however abstruse and complicated they appear, can be ultimately resolved into addition and subtraction.

It is principally then through the medium of the relations of quantity, that we are enabled to reason in a satisfactory manner, upon the objects of sense. For as every thing in nature, or what is the same thing to us, every sensation produced by one natural object, as compared with that produced by another, must be either equal, or similar; or unequal, or dissimilar : the whole are capable of being subjected, more or less perfectly, to the laws of quantity. This is effected in various ways, and by various artifices; but chiefly through the intervention of certain natural or assumed units, or standards of resemblance, as a second in time, a foot in space, &c.; and in proportion to the definite character of these units, or standards, or to their more or less satisfactory application; will the resulting branch of knowledge be more or less of a mathematical character; or be more or less rational and perfect.

By contemplating in the abstract, the boundless relations of time and of space, where no end can be conceived to addition and subtraction, we arrive at the only notions of infinity of which our nature seems capable. These once obtained, the obvious and necessary existence of cause, within the narrow sphere of our observation, naturally leads us to inquire, can this cause be infinite? And thus we are led by degrees, but irresistibly, to the sublimest of all conclusions; that a Cause or Agent, in every way commensurate with infinity—omniscient and omnipresent, eternal and omnipotent must exist—in other words, a God.

Compared, however, with infinity, and even with the objects of nature as they visibly exist around us, our actual knowledge of time and of space, is exceedingly limited. Like travellers on an extended plain, we see what is going on around us at the present moment; but the distant and the very near, the past and the future, are alike unknown to us. A few millions of miles, for example, or a few thousand years, comprise the utmost we know of space and of time. On the other hand, beyond the fraction of an inch, or of a second, every thing belonging to space and to time is inappreciable by our senses. Yet beyond these limits, we *know* that myriads of portions of space and of time must exist, too vast or too minute to be referred to our imperfect standards. Let us, for instance, take the distance of the nearest fixed star. This distance, we are assured by astronomers, is so great, that the utmost measure we can apply to it—the diameter of the earth's orbit—a space of no less than 192 millions of

miles is absolutely too little to measure it by—
is in fact, contained within it so many times,
that the number cannot at present be counted!
On the other hand, we shall presently find, that
the molecules of matter of which the objects we
see around us are composed, are so minute, that
a measure scarcely appreciable by the unassisted
sight—the thousandth part of an inch, for example—is vastly too large to compare them with,
and may in fact comprise millions of them!

Experience, the great and ultimate source of
all the knowledge we possess of those portions of
nature, to which our senses, and faculties, are
limited, may be acquired in two ways; by simple observation, and by experiment; that is to
say, either "by noticing facts as they occur,
without any attempt to influence the frequency
of their occurrence, or to vary the circumstances
under which they occur;" or, "by putting in
action causes and agents over which we have
control, and purposely varying these combinations, and noticing what effects take place."*
Now in all the higher departments of knowledge, the objects of which are principally
matter, and its motions in the aggregate; the
information we can acquire by one or both
these means is so complete, and at the same

* Herschel's Discourse on the Study of Natural Philosophy, p. 76.

time so favourable to the application of the relations of quantity, that the resulting sciences have all the certainty of abstract truths themselves. But when the knowledge we possess of objects is wholly sensible, and in no way commensurate, or only very imperfectly so, with their quantity, here it is that uncertainty begins; for though we may be able to trace the apparent *cause* and *effect* of a particular phenomenon; the most minute and careful observation and experiment, often give us but little insight into the connexion between the two, and generally fail us altogether. The origin of this failure is to be sought for, in the limited extent of our faculties; and particularly, in our complete ignorance of the nature of that mysterious communication, which we maintain with the external world, through the medium of sensation. In two of the senses indeed, *seeing* and *hearing*, we are able to trace the intermediate train of phenomena, between the external object producing the sensation, and the sensation itself, and even to form some idea of the remote cause of the sensation; but in the other two senses, *tasting* and *smelling*, the whole is involved in mystery from beginning to end.

Thus, when a bell is struck, philosophers have satisfactorily demonstrated that a vibratory motion, excited in the bell, and depending upon its elasticity, is communicated to the air in contact with it, and through this medium is propagated to the ear; in which organ, we know not why, the sensation of sound is excited. Circumstances very similar have been supposed to take place with respect to light; and *undulæ*, (or something obeying the laws of *undulæ*), have been demonstrated to exist, and to be propagated from the luminous body to the eye; thus the remote cause of sound, and probably of light, is proved to be motion. But in the cases of *tasting* and *smelling* the circumstances are altogether dissimilar; here the sapid and odoriferous matters are brought at once into actual contact with the sentient organs, and the sensations are the consequence, without any intermediate train of phenomena; at least any, that we can appreciate. What it is, therefore, in an acid or a rose, for example, analogous to motion in the bell, which produces the sensations we call *sour* and *sweet*, we know not, and probably never shall know; because the laws and relations of quantity are here either totally inapplicable, or can be only indirectly, and most imperfectly applied.

These observations are principally introduced with reference to the department of knowledge we have at present to consider. Almost all of what are denominated the *Chemical* properties of bodies, are objects of taste and of smell, rather than of sight and of hearing. Hence

P.　　　　　　C

they admit only of the indirect application of
the laws of quantity; and are the result, not of
reason, but solely of experience. Indeed, so
much is chemistry the creature of actual experimental research, that its simplest truths have
seldom been anticipated *a priori*. Thousands
of years of observation and experience, for
example, had not taught mankind that water is
composed of two elementary gaseous principles;
much less the proportions in which those principles combine to form water. Nay, even now
the fact has been established upon the clearest
evidence, we are unable to explain why it is so,
or even to comprehend the nature of the union,
or its result. In all chemical operations, therefore, (to adopt the language of Paley),—" our
situation is precisely like that of an unmechanical
looker-on, who stands by a machine, as a corn-
mill, a carding-machine, or a threshing-machine,
the fabric of which is hidden from his sight by
the outside case; or if seen, would be too complicated for his uninformed understanding to
comprehend. And what," continues this energetic writer, " is that situation? Ignorant as
he is, he does not fail to see that certain materials, in passing through the machine, undergo
remarkable changes; and what is more, changes
manifestly adapted for future use. Is it necessary that this man, in order to be convinced
that design, that intention, that contrivance,

have been employed about the machine, should be allowed to pull it to pieces to study its construction? He may indeed wish to do this for many reasons; but for all the purposes of ascertaining the existence of counsel and design in the formation of the machine, he wants no such intromission or privity. What he sees is sufficient. The effect upon the material, the change produced in it, the utility of that change for further applications abundantly testify, be the concealed part of the machine, or of its construction what it will, the hand and agency of a contriver."*

We have thus attempted to point out the rank which chemistry holds among the departments of human knowledge, and the kind of evidence which it furnishes in favour of design : the whole argument may be briefly recapitulated as follows :—chemistry is a department of knowledge founded solely on experience, for the phenomena of which we can assign no reason. But although the intimate nature of its changes be unknown to us, we see them manifestly directed to certain ends; hence, as objects directed to certain ends, where the whole of the intermediate phenomena can be traced and understood, always imply design ; we naturally infer design in others obviously so directed, even although

* Natural Theology, chap. vii. condensed slightly, but the argument strictly adhered to.

we may not be able to understand their intimate nature.

Such is the state in which Paley has left the argument; and while we admit that, even in its most perfect form, it is less satisfactory than that founded upon mechanism; we have always thought that our excellent author has not made quite so much of his subject as he might have done; and that the very imperfections and difficulties of chemistry, and of the allied branches of knowledge, give them some advantages over mechanism itself. When a series of wheels or of levers are arranged in a certain order, they *must* move in a certain way, and produce a certain effect, which can be foretold exactly. In such a case, we may admire the skill and ingenuity of the Contriver, or perhaps feel astonished at his power; but we scarcely do more: for much of the effect is lost in the apparent necessity of the result; and the consciousness that, under the circumstances, nothing else could have happened. When the Deity, therefore, operates through the medium of mechanism, He appears almost too obviously to limit his powers within the trammels of necessity; but when He operates through the medium of chemistry, the laws of which are less obvious, and indeed for the most part unknown to us; his operations, partaking more of the character of those of a free agent, appear of a higher order, and are more striking and wonderful. Do not, for instance, those extraordinary and mysterious changes constantly going on around us, beneath us, within us, derive no small additional interest from the very circumstance of their not being understood? Just such an interest, to revert to the argument of Paley, as the unmechanical looker-on feels in the operations of a corn mill, a carding machine, or a threshing machine; and to which he who is well acquainted with the mechanism, is a stranger? Certainly this is the case. Obvious mechanism, though well suited to display the intelligence and design of the Contriver, is not always so well adapted for arresting the attention of the observer; its very obviousness, in some measure, depriving it of its interest. But when we see the same Contriver, besides the most beautiful and complicated mechanism, employ other means utterly above *our* comprehension, though evidently most familiar to *Him*; the employment of these means is not only calculated to arrest our attention more forcibly, but at the same time to impress us with more exalted notions of his wisdom and power.

There yet remain one or two other points to be briefly considered, before we proceed to our subject. In the first place it may be asked, do those extraordinary changes which appear to be constantly going on in bodies around us, indicate real and substantial changes in the bodies

themselves; or are they mere phantasms and
creations of the organs of sense, through which
we become acquainted with them? The discussion of this question will probably be considered by most as superfluous; but for the sake
of those (if there be any) who entertain doubts
upon the point, it may be remarked, that the
sensations, though admitted to be mere signals
or indications, bearing little or no analogy with
the causes producing them, and therefore throw-
ing little light upon their nature, do nevertheless
represent real and substantial operations of some
sort, in the bodies themselves. This might be
proved, were it necessary, by a variety of arguments; but perhaps one of the most striking
arguments in favour of the reality of chemical
changes, may be deduced from the subserviency
to them, of those mechanical contrivances and
operations, every where existing in organized
beings. At least, half the mechanism in a living
animal is subservient to the chemical changes
constantly going on in it, and necessary to its
existence. Take, for instance, the circulation of
the blood: what a complicated apparatus is here
employed for the simple purpose of exposing the
blood to the action of the air in the lungs, in
order that it may there undergo some chemical
change. Now, surely no one can reasonably
doubt that this chemical change is as much a
reality, as the mechanism by which it has been

accomplished ; and if *one* chemical change be admitted to be a reality, why may not all others?

Finally, if there be any one who denies the existence of design, and sees nothing in all the more obvious arrangements and order around him, but the necessary results of what he chooses to denominate " the laws of nature;" let him calmly and deliberately consider the facts brought forward in the following pages : and if he can witness unconvinced all the numerous instances of prospective arrangement obviously made with reference to things not yet in existence; all the beautiful adjustments and adaptations of noxious and conflicting elements most wonderfully conspiring together for good ; and, lastly, the subversion of even his favourite " laws of nature" themselves, when a particular purpose requires it : if, we say, he can witness all these, and still remain incredulous of the evidences of design ; we can only observe, that his mind must be most singularly constituted, and apparently beyond the reach of conviction.

CHAPTER I.

OF THE MUTUAL OPERATION OF PHYSICAL AGENTS, AND OF MATTER; AND OF THE LAWS WHICH THEY OBEY.

" GOD has been pleased to prescribe limits to his own power, and to work his ends within those limits. The general laws of matter have perhaps the nature of these limits; its *inertia*, its reaction, the laws which govern the communication of motion, of light, of heat, of magnetism and electricity, and probably of others yet undiscovered. These are general laws, and when a particular purpose is to be effected, it is not by making them wind and bend and yield to the occasion, (for nature with great steadiness adheres to, and supports them,) but it is, as we have seen in the eye, by the interposition of an apparatus corresponding with these laws, and suited to the exigency which results from them, that the purpose is at length attained. As we have said, therefore, God prescribes limits to his power, that he may let in the exercise, and thereby exhibit demon-

strations of his wisdom.   For then, *i. e.* such
laws and limitations being laid down, it is as
though one Being should have fixed certain
rules ; and, if we may so speak, provided certain materials ; and afterwards have committed
to another Being, out of these materials, and
in subordination to these rules, the task of
drawing forth a creation : a supposition which
evidently leaves room, and induces indeed a
necessity for contrivance.   Nay, there may be
many such agents, and many ranks of these."*
This admirable passage from Paley is so much
in point, and so exactly expresses our opinions
regarding physical agents, that, as in a former
instance, we have chosen it as a text for illustration.   We shall proceed, therefore, to take a
summary view of " the limits within which the
Deity has confined his operations ;" that is to
say, of the laws by which matter, and those
subordinate agents by which matter is capable
of being influenced, have been made to mutually
act, and react upon each other.

The principles of activity, or *forces*, which
operate as subordinate agencies throughout nature, may be considered as of two kinds ; those
agencies which operate universally upon every
individual atom of matter, without reference to its
sensible properties ; as the forces producing the

* Natural Theology, chap. iii.

phenomena of *gravitation*, &c. :* and those agencies which operate among the different constituent molecules, of which all bodies are composed ;
and which are denominated *molecular*, or *polarizing* forces, &c. Of each of these subordinate
agencies, we shall in the first place endeavour to
convey some idea to the general reader.

* Many objections have been offered to the term *vis inertiæ*
adopted by Newton. Indeed, to speak of mere *inertia*, or *inactivity*, as a force, is obviously absurd. We have always agreed
with those who think that the term *inertia* has been unfortunately
chosen ; since *inertia* expresses only one quality, as it were, of
that which *is attracted*, or which *reacts*, in nature. But, we
fully acquiesce in the opinion, that whatever *resists attraction* or
reacts, is as appropriately named a force, in a certain sense of
that term, as that which *attracts* or *acts ;* and such resistance is,
in all instances, virtually considered as a force by the mathematician, however he may choose to designate it. Hence, for the
sake of analogy with what follows, we have adopted the supposition of two antagonist forces, viz. *inertia*, (for want of a better
name), and *attraction ;* which we have denominated the *forces*
of gravitation.

CHAPTER II.

OF THE INERTIA AND ACTIVITY OF MATTER.

To form a notion of what is termed the *inertia* or inactivity of matter; let us imagine a portion of it, as, for example, a ball of lead A, detached from all other matter, and existing absolutely uninfluenced in space. Such a mass of matter, if supposed to be at rest, must obviously remain so, for it cannot move itself; on the other hand, if it be supposed to be in motion, it must continue in motion; for it cannot be conceived to be able to stop itself, any more than it could be conceived to be able to set itself in motion: in short, a mass of matter under such circumstances of isolation, must be considered as perfectly passive and unable to change its state, whatever that may happen to be, whether of motion or of rest. Now let us suppose another portion of matter, as, for example, another ball of lead B, exactly of the same size as A, placed in free space at any moderate distance from A, and away from all other influences; what will happen? General experience teaches us, that under these circumstances, the two balls will mutually approach each other with an equal, but accelerated motion, till they meet at a point, exactly intermediate to those at which they first started; and the inference from this experience is, that the two balls exert a mutual and equal attractive force, which causes them to move towards each other. If the ball B, be twice the size of the ball A, the two balls will mutually approach each other as before; but, in this instance, instead of moving with equal velocity, while the ball A, moves two feet, the ball B, will only move one foot; or taking an extreme case, and supposing the ball B, to be indefinitely larger, say a million times larger, than the ball A, they will mutually influence and mutually move towards each other as before; but the motion of the ball B, will be so minute as to be insensible, while that of the ball A, will be the greatest possible. Here we have instances of the *inertia*, (inactivity, opposing force, &c.) and of the activity, (force of attraction, force of gravitation, &c.) which all matter exerts reciprocally, and mutually towards all other matter: and the laws of these forces, and the laws of the motions connected with them, as deducible from the circumstances stated, or from others, which it would be foreign to our present purpose to enter upon, may, in general terms, be given as follows:

" The mutual attraction of two bodies increases, in the same proportion as their masses are increased, and as the square of their distance is decreased; and, it decreases, in proportion as their masses are decreased, and as the square of their distance is increased."

These laws are absolutely general; and not only extend to the utmost limits of the universe hitherto explored by man; but to every form and condition of matter, without exception, and without reference to its other properties. They, therefore, constitute, probably, the most comprehensive " limits which the Deity has been pleased to prescribe to His power," and within which He operates with the most unceasing and undeviating regularity and certainty. They have also the remarkable property of being so amenable to the laws of quantity, or mathematics, as to be in most instances as firmly established upon reason, as abstract truths themselves. The mind of a Newton was chosen to reveal these laws to man; and man's acquaintance with them, may be justly considered as one of his noblest privileges. To point out their wonders in detail, and the sublime conclusions to which they lead, is the business of a colleague; at present we have to consider these laws in their more general form only, and, except in a single point of view, as objects of comparison merely, with those more immediately connected with our own subject.

The point of view to which we allude, is that peculiar case, or instance of gravitating force, termed *weight*. In our illustration of the attractive forces of matter above given, we supposed a case in which one ball was very much larger than the other: now this precisely corresponds with the case of the globe of the Earth, and of all common bodies near its surface. The Earth is more than 1,000,000,000,000,000 times the mass of any body which is observed to fall on its surface; and therefore, if even the largest body which can come under observation were to fall through a height of 500 feet, the corresponding motion of the Earth would be through a space less than the 1,000,000,000,000,000th part of 500 feet; which is less than the 100,000,000,000th part of an inch, and therefore quite inappreciable.* Now the attractive force exerted between the Earth and detached bodies, is denominated *weight*. Hence the weight of a body, at the earth's surface, is proportionate to its mass, or to the quantity of matter it may contain, whatever the form or qualities of that matter may be—a most important fact for the chemist; who, by employing the chemical properties of bodies as indications of identity or of change, is by these means enabled to apply to them the more certain measure of *weight;* and thus in some degree, to bring them under the dominion of the laws of quantity.

* Lardner's Cabinet Cyclopædia, Art. Mechanics, p. 79.

CHAPTER III.

OF MOLECULAR, OR POLARIZING FORCES, ETC.

In all chemical operations, as already observed, we only witness the beginning and the end ; the cause and the effect ; while the whole of the intermediate changes elude our senses. Nevertheless, by a careful observation of the phenomena we are enabled to form some notion of these operations ; and that amply sufficient to convince us of their wonderful nature. With a view therefore of arresting the attention of readers who may be unconscious of these wonders, or too apt to overlook them ; we have thought it proper to premise a sketch of what may be supposed to take place, among the ultimate particles of which all bodies are constituted ; during those remarkable changes which they are constantly undergoing. And here it may be remarked, once for all, that many of the views commonly entertained on these points, have always appeared to us to be so imperfect and unsatisfactory ; that so far from elucidating the subject, they have only served to render it the more obscure. In the following sketch, therefore, as better adapted for our purpose, we have endeavoured

to give that view of the subject, which, after
twenty years of close attention and no ordinary
labour, we have been induced to consider as
the most simple and consistent with the phenomena. The general reader, who feels no
interest in such enquiries, but who at the same
time wishes to be apprized of the nature of
the arguments deducible from the divisibility
and molecular constitution of matter, is referred
to the end of the present, and of the following
chapters, for a summary of these arguments.

Section I.

Of the Divisibility of Matter.

THE first point which naturally claims our attention in the consideration of molecular operations,
is the *size* of these molecules ; a subject usually
discussed under the head of the divisibility of
matter. Matter, or rather space, may be conceived to be divisible *ad infinitum ;* at least no
limits can be assigned beyond which its subdivision cannot necessarily proceed. But, there
cannot be the least doubt that matter, as it
exists in the world around us, is composed of
ultimate particles or molecules, incapable of
further division or change ; at least, by ordinary

agents: the reasons for these assertions will appear hereafter; at present it is our object to convey to the general reader some idea of the magnitude of these particles, with the view, prin cipally, of showing how infinitely they surpass the limited powers, not only of our senses, but almost of our conception. The subject, however, has so much attracted the attention of philosophers, that most of our readers must be already familiar with it; we shall therefore content ourselves with merely selecting a single instance from each of the kingdoms of nature.

As an instance from the mineral kingdom, we may quote from Dr. Thomson, who has shown that an ultimate molecule of lead cannot weigh more than the 1-310,000,000,000th, nor an ultimate molecule of sulphur, more than the 1-2,015,000,000,000th of a grain, and probably a great deal less; and that the *size* of the molecule of lead cannot surpass, and is probably much smaller, than the 1-888,492,000,000,000th of a cubic inch!* The vegetable kingdom presents us with innumerable instances, not only of the extraordinary divisibility of matter, but of its activity, in the almost incredibly rapid developement of cellular structure in certain plants. Thus the *Bovista giganteum* (a species of fungus), has been known to acquire the size of a

* System of Inorganic Chemistry, I. 7.

P.   D

gourd in one night. Now, supposing with Professor Lindley, that the cellules of this plant are not less than the 1-200th of an inch in diameter, a plant of the above size will contain no less than 47·000·000·000 cellules ; so that, supposing it to have grown in the course of twelve hours, its cellules must have been developed at the rate of nearly 4·000·000·000 per hour, or of more than sixty six millions in a minute !* and when we consider that every one of these cellules must be composed of innumerable molecules, each one of which is again composed of others ; we are perfectly overwhelmed with the minuteness, and number of the parts, employed in this single production of nature. But the animal kingdom perhaps presents us with still more striking instances than these. Thus animalcules have been discovered whose magnitude is such, that a million of them do not exceed a grain of sand ; and yet each of these creatures is composed of members as curiously organized as those of the largest species. They have life and spontaneous motion, and are endowed with feeling and instinct ; in the liquids in which they live, they are observed to move with astonishing speed and activity ; nor are their motions blind and fortuitous, but evidently governed by choice, and directed to an end. They use food and drink,

* Introduction to Botany, p. 7.

from which they derive nutrition ; and are therefore provided with a digestive apparatus. They have great muscular power, and are provided with limbs and muscles of strength and flexibility. They are susceptible of the same appetites, and obnoxious to the same passions, as the largest animals. Must we not conclude that these creatures have hearts, arteries, veins, muscles, sinews, nerves, circulating fluids, and all the concomitant apparatus of a living organized body? and if so, how inconceivably minute must these parts be? If a globule of their blood bears the same proportion to their whole bulk, as a globule of our blood bears to our magnitude, what power of calculation can give an adequate notion of its minuteness?*

We have thus endeavoured to convey some conception of the magnitude of the ultimate molecules of which bodies are composed; but though we have succeeded in showing that they cannot exceed a certain magnitude; we are by no means certain that they are not in reality much less—indeed a great deal less, than the least magnitude of which we have endeavoured above, to convey a conception: yet notwithstanding this inapproachable minuteness, the ultimate molecules of which bodies are composed, retain, in the highest degree, all the

* Lardner's Cyclopædia, Art. Mechanics, p. 13.

characters of matter; and moreover, possess certain remarkable properties in common, upon the nature of which, we shall, in the next place, make a few remarks.

Section II.

Of the Forms of Aggregation of the ultimate Molecules of Matter.

Matter in the aggregate, and as it appears to exist in the world around us, is known to us principally in three forms or conditions:—the *Solid*, the *Liquid*, and the *Gaseous* (the latter including the state of vapour, and the etheriform condition of matter). These three forms or conditions of matter, in their well marked states are sufficiently distinct; though the whole gradually run into each other—the solid into the liquid, and the liquid into the gaseous, by such imperceptible grades, that in many instances, it is not easy to say, where one ends, and the other begins.* The notions, which the mechanician or natural philosopher employs in reasoning on these forms

* It may be proper here to observe that some bodies, as water, for instance, are capable of existing in that imperfectly gaseous form denominated *vapour*, under all ordinary circumstances; thus even ice gives off vapour rapidly, as we shall find hereafter.

of bodies are—of a solid, that all its parts are indissolubly and unalterably connected, and impenetrable, so that the relative situation of the parts among one another, cannot be changed, or one part be set in motion without all the rest; of a liquid, that all its parts are freely moveable among one another, but that it is not dilatable or compressible by mechanical means; of a gas or aeriform body, that all its parts are not only freely moveable among one another, but that it is compressible and dilatable without limits. Strictly speaking, however, there are no objects actually existing in nature which completely conform to these definitions: no solid, for instance, absolutely hard and impenetrable; no fluid not compressible and dilatable; no gas-compressible or dilatable without limits: and these circumstances are evidently the necessary result of all the objects in nature being composed of aggregations of the minute molecules we have been considering. Thus solids composed of such molecules, must necessarily have innumerable interstices or pores; hence, when submitted to pressure, they always undergo more or less of condensation, and apparently occupy less space than before: the same remarks may be made with respect to liquids; while gaseous bodies, supposed to be composed of such molecules, cannot, of course, be infinitely compressible.

SECTION III.

Of the solid Form of Bodies.   Crystallization.

NATURAL solids present us with a variety of properties usually termed *secondary*, many of which
are of the utmost importance : such are *hardness*
and *softness, elasticity, toughness, malleability,
tenacity, ductility*, &c., all too well understood to
require definition here.   These properties evidently depend in a great degree upon original
differences in the properties of the component
molecules themselves; but there is no doubt that
many of them are also intimately connected
with the modes in which the molecules are arranged.   Of these modes we can form no precise
idea in a great many instances ; there is, however, *one* form of solid aggregation, the *regular
crystalline form*, which has occupied much more
attention than the rest, and upon this form we
proceed to offer a few remarks.

As an object of illustration we shall select the
familiar one of *water ;* which from its well known
properties of existing either as a solid, a liquid,
a vapour, or a gas, by a slight variation of circumstances, is well adapted for our purpose; as
we are thus enabled to employ the same object
of illustration throughout.   At present we have
to consider water in its solid form of *ice.*

Every one must have remarked that water, in the act of freezing, assumes various symmetrical forms, shooting into spiculæ, &c., as may be beautifully seen on our windows on a frosty morning. Now this affords a familiar instance of what is termed *crystallization;* a property apparently possessed by all ponderable matter, and readily exhibited by such matter, when under favourable circumstances: and it has been remarked, that the form assumed by the same matter is usually similar, or easily deducible from some common form, according to well-ascertained and obvious laws. Let us now briefly inquire into the properties which the ultimate molecules of water must be supposed to possess, to enable them to form these symmetrical aggregations.

In the first place, it is evident, that the simple supposition of mutually attractive forces between these molecules, analogous to, or identical with, the forces of gravitation, is inadequate to explain the phenomena. Possessed of such properties alone, the ultimate molecules of bodies might indeed be imagined to adhere together, and their aggregations might even exhibit something like regularity; but this regularity would in a great measure be accidental, and probably never twice alike: hence the utmost latitude of assumption would never enable us to explain upon such principles alone, that sameness of figure above alluded to, as always assumed by

the same matter. It is obvious, therefore, that
the ultimate molecules of bodies are influenced
by other powers than those of simple *inertia* and
attraction : what is the nature of these powers?
On this point there have been various opinions.
Some have supposed the ultimate particles of
bodies to possess shapes identical with those of
the aggregates which they form ; that a crystal,
for example, whose shape is a cube, is formed
by the aggregation of a number of infinitely
little cubes, &c.   But to others this supposition
has appeared so improbable, and so unlike the
usual simplicity of nature's operations, that
they have rejected it, and have formed the more
feasible hypothesis, that the ultimate molecules
of bodies are either spheres or spheroidal ; that
is to say, more or less, *virtually* globular.*

Let us take it for granted then, that the ultimate molecules of bodies are spheres : with
what powers is it necessary to suppose these
little spheres to be invested, in order to enable

* Strictly speaking, perhaps this observation is applicable to
the forms supposed to be assumed by the influences surrounding
the molecules, and by which all their operations are directed,
rather than to the absolute forms of the molecules themselves;
which, though in all instances virtually exerting spheroidal influences, must, in different instances, have very different forms.
Those who wish to study the principles upon which spheroidal
molecules may be supposed to aggregate into crystalline forms,
are referred to Dr. Wollaston's interesting paper on the subject
in the Philos. Trans. 1813, p. 51.   It may be noticed, however,
that the principles we shall advance differ materially, in other
respects, from those referred to.

them to cohere, and to form the symmetrical figures we observe among natural bodies? The existence of simple, mutual, and general attractive powers among such a set of molecules, has been already observed to be inadequate to explain the phenomena; there must be some specific powers, determining similar particles to combine in similar ways, otherwise the same resulting forms could not be supposed to be produced. In the three small spheres,* Fig. 1, let us suppose the points *E*, *E*, *E*, and *e*, *e*, *e*, on their superficies, to be endowed with the following properties; viz. that the similar points, *E*, and *E*, upon any two of the spheres, have the property of mutually repelling each other; while the dissimilar points, *E*, and *e*, upon any two of the spheres, have the property of mutual attraction. In such a case, the three molecules will readily combine *E*, to *e*, as in Fig. 2, but in no other way. Now let us suppose the same three spheres to be endowed with properties at the points *M*, *M*, *M*, and *m*, *m*, *m*, as in Fig. 3, nearly resembling the properties with which they are endowed, at the points *E*, *E*, *E*,

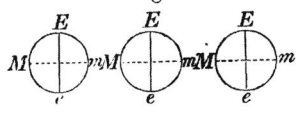

Fig.3.

---

* Or rather sections of spheres, and the same is to be understood of all the subsequent figures.

and *e, e, e.* Spheres so endowed will aggregate readily, as in Fig. 4, *E*, to *e*, and *M*, to *m*, but in no other way; and thus instead of a single line, we obtain a plate of molecules, one in thickness.* To form the third dimension, or to constitute a solid; it is necessary to assume the molecules as in Fig. 5, to be possessed not only of the

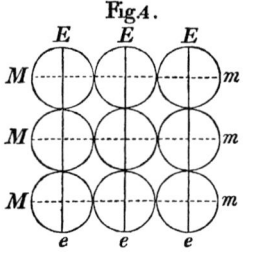

Fig. 4.

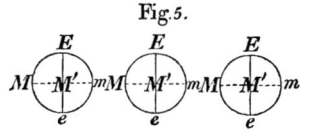

Fig. 5.

attractive points *E, E, E,* and *e, e, e, M, M, M,* and *m, m, m,* but also of the attractive points *M′, M′, M′,* and *m′, m′, m′,* (the point *m′* being supposed to be opposite to the point *M′,* and out of sight). Molecules so endowed will readily combine as in Fig. 6, and form a cube, or some figure obviously deducible from it, but in no other manner: and in this

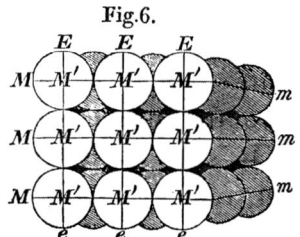

Fig. 6.

ble from it, but in no other manner: and in this

* Here it is to be observed that the similar poles *E, E,* and *e, e,* of each pair of molecules being supposed to be repellent within certain limits, as will hereafter be explained; their absolute contact is prevented, and the two molecules are balanced, as it were, between the two opposing and the two attracting forces. The consideration of forces operating in these, and in the other modes, subsequently mentioned, present some highly interesting and novel objects of research for the mathematician.

way, by assuming certain attractive and repulsive points upon our spheres, at appropriate parts of their superficies, it is not difficult to conceive them capable, in different instances, of forming aggregates of any shape whatever. The question to be next considered is, how far are we authorized in making such apparently complicated and gratuitous assumptions respecting the properties of the ultimate molecules of matter; are there any phenomena in nature justifying such conclusions, and what are they? And this leads us to enquire further, but as briefly as possible, into the phenomena of aggregation, as we see them constantly going on around us.

Aggregation is usually considered to be of two distinct kinds; viz. aggregation depending on the simple cohesion of *similar* molecules of matter, as of water; which similar molecules for the present, may be supposed to undergo no change by the combination: and aggregation depending on the union of *dissimilar* molecules of matter, capable of exerting a mutual *chemical* change upon each other: in which kind of aggregation, the aggregate produced is a *tertium quid,* or third something, differing altogether from either of the original molecules composing it. Now both these conditions of aggregation obviously exist in the same substance; at least when in the solid form. In the present case of water, chemical aggregation is, in the first place, exerted between the heterogeneous molecules, hydrogen and oxygen, which uniting, form compound homogeneous molecules, (of water); while in the next place, the molecules of water uniting *chemically* in one direction, and *cohesively* in the other, form the solid crystal (of ice). Thus chemical aggregation, and cohesive aggregation, are as distinct as the polarities themselves upon which they depend ; and if the one kind of aggregation existed alone without the other, no such thing as a regular crystalline solid would probably be formed in nature.

From the above views, of molecular forces, it follows as a consequence, that every molecule must possess, in one axis, powers and properties totally different from those with which the molecule (or molecular aggregate) is endowed, in the other two diameters or axes. This axis and its polarities, may, by way of distinction, be called the *chemical* axis and polarities ; and may be supposed to be represented by the axis and polarities *E, e*, in the preceding figures. The other two diameters, (and indeed every other diameter that can be supposed to be drawn, through the centre, from opposite points of the superficies of the molecule), probably possess common properties, and may be called the *Cohesive* diameters and polarities. Here then the existence of two forces, is indicated,—the one axial, the other equatorial, if we may be allowed the expressions. The next question is, do forces actually exist in nature on the large scale, which are thus related to each other, and what are these forces? Now late observations have proved, beyond a doubt, that the *Electric* and *Magnetic* forces are so related to each other. We proceed, therefore, to take a short view of Electricity and Magnetism.*

Electricity.—It would be foreign to our present purpose, to enter into details respecting this, and other departments of science, to which we may have occasion to allude: we shall therefore content ourselves with a short summary of their general principles. It seems to be generally admitted, that the phenomena of electricity depend upon two energies, usually existing throughout nature in a state of equilibrium, in which state their peculiar powers are not perceptible; that this equilibrium is capable of being destroyed by a variety of causes, as friction, &c. ;

---

* It may be remarked, that as all parts of the superficies of our molecules, except the *chemical* poles, are supposed to be more or less capable of cohesion, their aggregation in the form of common crystallized solids may be readily conceived. With respect to the cohesion (if we may be allowed the expression) of the different chemical poles E and e, of similar molecules with each other, such cohesion seems to be proved by several circumstances, which it would be foreign to our purpose at present to inquire into; but of which, perhaps, the optical properties of crystals will hereafter form one of the most striking illustrations.

and that owing to the different capacities possessed by different bodies, for conducting and
retaining the electric energies, these energies can
be partially separated and kept asunder; in
which state, they are capable of exhibiting their
peculiar powers. The powers thus exhibited
are such, that if two bodies, charged in excess
with the same energy, be brought into the vicinity
of each other, they mutually repel each other;
while two bodies, charged with the two different
energies, mutually attract each other. In this
disturbance of the equilibrium of the two energies, it is to be remarked, that in no instance
do we suppose, that the two energies are, or can
be, entirely separated, so as to reside each, *per se*,
in different bodies; but that a portion of the
energy of the one body goes to the other body,
which at the same time returns a corresponding
portion of its antagonist energy; hence, other
things being equal, each body contains the same
total quantity of the two electricities, as before
the equilibrium was destroyed.

Such are, we believe, the general opinions
respecting the fundamental laws of action and
equilibrium of the two electric energies. There
are certain phenomena immediately arising out
of them, which, as they are the most frequent
and important of all those connected with the
disturbance of the equilibrium of the two
energies in different bodies, we shall briefly ex-

plain : we allude to what are usually called the phenomena of *induction.* Suppose an electrified body A, (that is to say, a body having the equilibrium of its electric energies destroyed, as above mentioned,) be brought into the neighbourhood of another body, B, in its natural state ; what takes place? The electricity E, of the body A, acting upon the corresponding electricity E, in the body B, repels this electricity, E, to the other end of the body B, which is furthest from the body A : at the same time, the other and opposite electricity, *e,* is attracted to the end of the body, B, which is nearest to the body A. The body B, therefore, while under the influence of the body A, will exhibit all the phenomena of electricity, and is said to be *electrified by induction;* but if the body A, be removed from the neighbourhood of the body B, immediately the natural equilibrium of the energies of the body B, will be restored, and all signs of electricity will vanish. In this experiment neither of the bodies gains or loses any thing. As these phenomena are constantly occurring in nature, and as we shall have occasion to use the term *induction,* we have endeavoured to give an explanation of the phenomena, intelligible to the general reader.

Of Galvanism.—While we are upon the subject of electricity, we may briefly notice that important modification of it termed *Galvanism.*

This form of electricity, instead of being evolved by friction, is usually obtained by the mutual action of various metals, and chemical agents upon each other. Late experiments, however, have shown that the energies thus developed, differ in no respect from those of common electricity ; but that they are obtained in this way in much greater *quantity* only, though in a lower state of *intensity*, than by the common machine ; and that many of the supposed peculiar effects of galvanism, are the consequences of the *motion*, of such large quantities of these energies, through bodies of various conducting powers. Galvanism has recently attracted much more attention than ordinary electricity, from the facility with which it may be applied to the purposes of the chemist ; and from the extraordinary light it has thrown upon many chemical phenomena. Indeed, the chemist has been more indebted to the energies of galvanism than to any other ; and he will probably be still further indebted to them than he yet has been. The phenomena of galvanism, in most respects, so closely resemble the phenomena of electricity, that they do not require further illustration here.

Of Magnetism.—The general phenomena and laws of magnetism are very analogous to those of electricity. There are evidently two antagonist energies, which, while in a state of equilibrium, are not cognizable; but when separated, each one is mutually repellent of its similar, and mutually attractive of its opposite or antagonist. Thus the two north or two south poles of two magnetic needles mutually repel each other; but the north pole of one needle and the south pole of another, mutually attract each other. Bodies are also rendered magnetic by induction, when in the vicinity of another magnet, precisely as happens with respect to electricity. Magnetism principally differs from electricity, in being apparently limited to a few bodies, as iron, and two or three others; but late observations have thrown an entirely new light on this part of the subject, which we have next to consider. Before we proceed, however, we may make a few remarks upon the obvious questions :—

What becomes of the two electric and two magnetic energies when in a state of equilibrium? Do the electric and magnetic energies combine to yield the same, or a different result? and what is the nature of this result or results : and in what form do they exist around us? On these questions different opinions have been held: some supposing that both the electric and the magnetic energies, when in a state of equilibrium, alike constitute *heat*; others, something else. That both are most intimately connected

P.   E

with heat and light, is evident; but at present
we decline to give a decided opinion on the
nature of the connexion.

We come now to enquire into the relations of
electricity and of magnetism to one another—a
discovery which we owe to Oersted, and one of
the most important that has been made in the
present age. The following is a summary of
Oersted's discovery. Let us suppose, in the
annexed figure, E e, to represent the
wire connecting the zinc and the
copper terminating plates of a common galvanic battery in action. From
what has been said, it may be conceived, that under these circum-
stances, there will be two currents *moving* through
this wire in opposite directions ; (from the copper to the zinc, usually called *positive* electricity;
and from the zinc to the copper, usually termed
negative electricity.* Now in this state of things,
it has been satisfactorily established by experiment, that besides these two currents, there are
two others having totally different properties,
indeed all the properties of the magnetic ener-

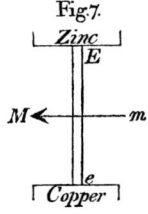

Fig.7.

* The reader will observe, that the above observations apply
to a wire connecting the terminating plates of a *common galvanic
battery ;* which plates are in reality superfluous. Hence the currents and polarities here given, are just the reverse of those which
actually exist, in a wire connecting the zinc and copper in a
simple galvanic circle. We have thought it proper to notice
this circumstance, to prevent misconception.

gies, *moving*, not in the direction of the wire, but *in circles, or rather spirals, round it.* The energy corresponding to the north pole of the magnetic needle moves from right to left, round the wire as above posited; while the energy corresponding to the south pole of the magnet moves in the opposite direction, or from left to right. Hence when a delicate magnetic needle, *M m*, is suspended *above* the wire *E e*, its north pole *M*, will be attracted by the current moving from left to right, with which it comes first in contact; and its south pole, for similar reasons, will be attracted by the opposite current. A needle so suspended will consequently assume the direction represented in the figure, with its north pole *M* to the left: and if it be carried round the wire by its point of suspension, it will be always found to keep the same relative position with respect to the wire. Thus when *below* the wire, the needle will apparently point in the opposite direction; when on the same level, on the *left* hand, vertically downwards; when on the *right*, upwards.

Bearing in mind these relative positions of the currents and needles; in what follows we may neglect the currents, and judge from the position of the needles alone. Let us consider the case of two connecting wires placed by the side of each other, as in the figures annexed; and which wires may be supposed to represent the chemical axes of our molecules. Now these wires, in consequence of the magnetic energies circu-

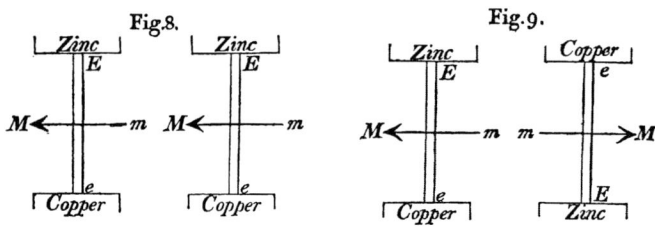

lating round them, will mutually attract or repel each other, according to their position. If, as in Fig. 8, they are both in the same relative position, they will mutually attract each other; as may be inferred from the position of the needles, M m, M m, the north pole of one of which corresponds with the south pole of the other; but if one of the wires be reversed, as in Fig. 9, they will mutually repel each other, the two similar poles of the needles in this case being contiguous. These relations hold universally, and what is most important, recent observations have shown them to be, under certain slight modifications, reciprocal; that is to say, if the magnetic energies be made to move in straight lines, the galvanic energies are found to circulate round them, nearly according to the laws above described, as happening to magnetism, round electricity. Hence electric sparks, and indeed all the phenomena of electricity, can now be obtained from a common magnet.

Whether electricity and magnetism be different forms of the same energies, resulting from the different direction of their motions; whether they be distinct energies; whether they be the cause, or the effect of *Polarity*; we shall not stay to inquire: it is sufficient for our present purpose to know, that they are inseparably associated with one another in the manner stated; and are always present, at least, in, if they be not the immediate cause of, all molecular actions among ponderable bodies. And this brings us back to the point, where we digressed to consider the subjects of electricity and magnetism.

We attempted to show, that the ultimate molecules of matter must be endowed with two kinds of Polarity; one which we have denominated *Chemical* Polarity, of a binary character, existing between molecule and molecule, and chiefly between molecules of *different* matter; another denominated *Cohesive* polarity, determining, under certain circumstances, the cohesion of the molecules of the *same* matter. We further attempted to explain how these polarities must exist, or be distributed in our molecules, so as to fulfil the offices assigned to them, and which they evidently fulfil in nature. Lastly, we have shown, that the electric and magnetic polarities, or energies are actually related to one another, precisely in the same way, that we supposed the chemical and cohesive polarities to be.

The question then at once arises,—are these forces identical? Do the electric polarities correspond with the supposed chemical polarities; and do the magnetic polarities correspond with the cohesive polarities of our molecule?

To us, we have no hesitation in saying, this conclusion seems very probable, nay, almost inevitable; not only for the reasons stated, but for others equally striking, which we shall have occasion to refer to hereafter. In the mean time we may briefly consider the subject with a little more attention, and principally with reference to some apparent objections, which may be raised against it. In the first place it may be objected, that it is difficult, from what we know of the varying and capricious character of the electric energies, to suppose they can ever exist in that definite and permanent form, in which they must exist, if they be really identical with the cause of chemical affinity. To this objection, it may be replied, that magnetism can, and does, exist permanently in bodies for ages; and as electricity is an inseparable attendant upon magnetism, this energy must also have equal permanence. Again, a portion of zinc and a portion of copper placed in contact, produce electrical effects as constant and enduring as the metals themselves. The argument, therefore, founded upon the want of permanence, and uniformity, of the electric and magnetic energies, cannot, if duly considered, be supposed to have any weight: for the molecule may be conceived to be composed of two parts analogous to the copper and zinc in contact; and the electricity and accompanying magnetism evolved, may be supposed to be as permanent in their character, as the parts of the molecule evolving them.   To the argument, that electricity and magnetism, as we are acquainted with these energies, are *inadequate* to produce the effects, and explain the phenomena, of chemical affinity and cohesion ; it may be replied, that they may be so ; but that these energies, as we are acquainted with them, are probably merely accidental and peculiar modifications of the real energies ; which in their elementary form, may be something altogether different, and quite unknown to us.   In proof of this notion, it may be observed, that the electricities of the common machine, and of the galvanic machine, apparently differ materially; while the electricity existing in certain animals appears to differ from both.   The magnetism evolved by electricity differs also slightly from common magnetism; yet no one now doubts that these differences arise from varieties in the quantity and intensity of the same energies; which in their elementary form, therefore, may, and probably do, differ from all these varieties.   At any rate, we are unable to say that one of these varieties is more elementary than another; and consequently we have no right to assume that either of them is elementary; much less to found any argument upon the assumption.

Before we quit the subject of polarities and polarizing forces, it may not be amiss, in the last place, to make a few general remarks on the properties, in which these forces resemble, or differ from, the forces of gravitation.

The forces of gravitation, *inertia* and attraction, appear to be associated, and to reside in every individual *atom* of matter in the universe; hence every atom mutually attracts, and is attracted, by every other atom. The polarizing forces, on the other hand, are evidently disassociated, and reside in different parts of the same *mass:* hence this mass can in no instance be a mathematical point, (or atom?), but must consist of at least two parts: hence, also, as all matter appears to possess polarity, matter must exist in the state of mass or *molecule;* each of which molecules must occupy actual space. Thus the forces of gravitation, and polarizing forces are quite distinct. The forces of gravitation are primordial, and probably co-existent with matter; while polarizing forces have more of a derivative or resultant character; and are evidently subordinate to those of gravitation. The question here naturally arises,—Are these different forces related to one another? Do polarizing forces consist of the forces of gravitation in a state of separation, (if we may be allowed the expression,) or do they result from the motion of the molecules upon their axes? Such questions are quite beyond our powers,—indeed we have nothing at present to do with them,— our object here, being merely to point out the apparent limits, within which the Deity has chosen to confine his operations.

SECTION IV.

Of the Liquid Form of Bodies. Of Heat.

HITHERTO we have spoken of the aggregation of molecules in the solid form only; we have next to consider their arrangement in that state in which they constitute a liquid. Our notion of a fluid, generally speaking, is, that all its parts, or molecules, instead of being fixed, are perfectly movable among one another; our notion of a liquid (the least perfect form of fluidity) is, that its molecules are not only movable, but incompressible. Now, still retaining water as our example of a liquid; let us consider what must happen to its molecules situated, as we supposed them to be, in the form of ice; before they can be so arranged as to constitute the

liquid water. A moment's reflection teaches us,
that they must be loosened or separated from
each other; and, as they cannot separate themselves, that some new agency is requisite for
this purpose. It need scarcely be mentioned,
that this agent is *Heat;* on the general phenomena and laws of which most important principle, we now proceed to make a few remarks.

Of Heat. The sensations termed *heat* and
cold are too well known to require explanation.
These sensations, however, like all others, are
merely the effects of some external cause or
causes, operating on, and through, our organs, in
a manner totally unknown to us. On that cause,
and the mode of its operation, various opinions
have been entertained: some considering the
cause of heat (caloric) to be an existent and material fluid; though of such extreme tenuity and
imponderability as to escape our observation,
and to become manifest to us only by its effects
upon our sensations, and upon all the ponderable forms of matter: others considering the
cause of heat to be not material, but a property
or principle of motion; which, by exciting a certain species of vibration among the particles of
bodies, gives rise to the sensation and effects of
heat. Such are the most usual opinions, and
the probability is, that they are neither of them
literally correct; but that heat, and we may add

light, are substances, the molecules of which are influenced by polarizing forces, precisely similar in all respects, to those which influence common matter; that is to say, that the molecules of heat and of light obey laws, similar in all respects to those laws which govern the molecules of ponderable bodies.* We have already alluded to the opinion maintained by some, that heat is a compound principle, consisting of the two forms of electricity in a state of equilibrium. We now draw the attention of the reader to this hypothesis, in order to state, that whatever heat may consist of besides, it is almost impossible to explain its effects upon the polarizing forces, without supposing that it at least involves, if it do not pass into, the electric forces, upon which the polarizing forces *appear* to us to depend. We have said *appear ;* for as has been already stated, though it be convenient to consider the polarizing forces under the forms of electricity and magnetism, in which they are most usually and palpably manifested to us among ponderable bodies ; yet, in their elementary form, these forces may in reality be something very dif-

---

* We are aware that this opinion is opposed to that of most mathematicians, who favour the undulatory theory of light; and with good reason, so far as they have occasion to consider it: but we are decidedly of opinion, that the *chemical* action of light can be explained only upon chemical principles, whatever these may be. Whether these chemical principles, will hereafter explain what is now so happily illustrated by undulæ, time must determine.

ferent; not only from those of electricity and
magnetism, but from all others with which we
are acquainted: while electricity and magnetism
themselves, as we know them, may be nothing
more, than the effects of these elementary forces,
upon the subtle matters, of which the electric
and magnetic molecules are composed.*

Of the Effects of Heat.—One of the most
general effects of heat, is the *increase of volume*
which it produces, in all bodies in which it
is accumulated. There are a few exceptions
to this law; and one of so important a character as to require especial notice hereafter.
In the mean time, however, the generality
of the law may be taken for granted; and its
mode of operation on the molecules of bodies,
may be illustrated in the following man-

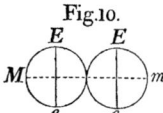

Fig.10.

ner : † As before, let us suppose
Fig. 10 to represent two molecules of ice, in which the chemical axes, *E e, E e*, are parallel,
and the similar poles of these axes are in the
same direction. In this position of the chemical
axes, the similar poles of these axes will be
repulsive; but their repulsive effects are supposed to be limited, and nearly quiescent.

* See Appendix.

† The general reader will observe, that the molecules of heat
are vastly less than those of any ponderable substance; otherwise
the effects ascribed to them, could not, of course, be imagined to
take place.

Hence it is presumed, that the repulsion of these poles does not prevent the molecules from cohering at their equators ; every point, *M, m,* of which, in the two molecules, as thus arranged, will be dissimilar, and, of course, attractive. Let us now suppose, from some external source, a certain quantity of heat to be communicated to these molecules. The natural tendency of heat is usually considered to be, to arrange

Fig.11.

itself in the form of an atmosphere, around the molecules as in Fig. 11 ; the consequence of which is, that while the temperature of the molecules is raised, they are, at the same time, slightly separated from each other : thus the molecules of ice to which heat has been applied, will be partially disjoined at the cohesive points, *M, m ;* and the result, of course, will be, that these molecules will occupy, as in Fig. 11, more space, than they occupied, as in Fig. 10, before the addition of the heat. The preceding explanation of the effects of heat in separating the molecules of solids, and thus of increasing their volume, applies also, (though perhaps in a still more striking degree,) to the effects of heat upon the molecules of bodies, in the liquid and gaseous states.

Another important property of heat, which we must briefly explain, is what is termed the *Latency* of heat. When the *same quantity* of heat is introduced into different bodies, they exhibit very different *apparent* temperatures. This property, as possessed by different bodies, constitutes what is called by chemists, their *capacity for heat*; or, their *specific heat*. Thus, if the same quantity of heat, which we supposed to have been introduced into two molecules of ice, had been introduced into two molecules of silver; the apparent temperature of the molecules of silver would have been raised upwards of ten times more, than the apparent temperature of the molecules of ice; hence, in the case of the ice, some of the heat must have disappeared, or become, in the language of chemists, *latent*. The latency of heat appears to depend upon two different causes; or rather, properly speaking, there are two distinct modes of the latency of heat; as may be thus illustrated. Let us take the two bodies above mentioned—ice and silver: these bodies, under the same volume, contain very unequal portions of matter; the silver being ten times as heavy as the ice. The vacuities in the ice therefore, must be very much greater than the vacuities in the silver; hence, when equal quantities of any matter capable of occupying such vacuities, as heat may be supposed to be, are introduced equally into both these bodies, very dissimilar apparent effects must ensue. The more porous body will absorb, and condense within its vacuities, the matter which had been added, and will leave little of that added matter external and sensible; while the less porous body, having less room among its pores, will condense less, and will exhibit externally, and sensibly, a larger quantity of the added matter. The more porous body, therefore, may be said to have a greater *capacity* for heat, than the less porous body; from its greater power of absorbing heat, and rendering it latent.* Such, we believe, is the usual explanation of the latency of heat in any body, comparatively with other bodies; and, to a certain extent, it is probably correct: but there is another mode of the latency of heat, apparently quite different from the above noticed; and which does not seem to admit of being explained on the same principle: this we have now to consider.

Let us suppose that into a mass of ice, which has been cooled to several degrees under the freezing point, a uniform and regular flow of heat be determined from some external source. In consequence of this accession of heat, the temperature (and volume?) of the ice will be

---

* This union of heat with ponderable bodies may, perhaps, be considered as analogous to the condensation of gaseous bodies by porous substances—a very remarkable set of phenomena, which deserve to be much more carefully studied than they have been. The absorption of light seems also to be of a similar nature. All these phenomena apparently depend upon chemical principles; and probably, if studied in connexion, would mutually illustrate each other.

gradually augmented; till the temperature be
raised again to the point of freezing. At this
point, the ice will begin to be thaw'ed, so as to
become water; but notwithstanding, heat continues to flow into the melting ice, its temperature will remain stationary, at the point of
freezing, till the whole of the ice be melted; to
accomplish which complete melting, a quantity
of heat, equal to 140 degrees of Fahrenheit's
thermometer, will be found to be necessary.
When the whole of the ice has been melted, if
the heat still continue to flow as before, the
water will acquire apparent temperature, in the
same manner, in which apparent temperature
had been acquired by the ice. Now this latency of heat flowing into melting ice, is produced by the real disappearance of a quantity of
heat, equal to 140 degrees of Fahrenheit's thermometer. But the latency of heat in melting
ice, does not admit of the same explanation as
the latency of heat in ice under the freezing
point; or in ice, comparatively with silver: the
water, instead of being greater in volume, and
consequently having greater vacuities, than
the ice, from which it was formed, is actually
less in volume, and must therefore contain fewer
vacuities. How, then, are the phenomena to be
explained? We commenced our observations
on heat, by alluding to the hypothesis, that under
certain circumstances, heat appears capable of

passing into two energies, if not identical with, at least operating in the same way, as those of electricity. We may suppose, therefore, that the 140 degrees of heat which disappear during the melting of the ice, are converted, in some unknown manner, into the two polarizing energies ; and that the energies, thus produced, are superadded to the energies already existing attached to the molecules of water, the total quantity, or intensity, of the molecular energies of which, are increased by such additions. The changes effected in the relative position of the chemical axes of the molecules of ice, during its conversion into water, by the increase of the polarizing energies arising from the added heat, may be thus illustrated : let Fig. 12, represent

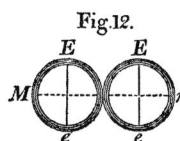

Fig.12.

two molecules of ice, having their chemical axes parallel, and the similar poles of these axes in the same direction, in which they were represented in Fig. 10 and Fig. 11. In such a position of the chemical axes, their similar poles E, E, and e, e, are of course mutually repellant, though in a low degree ; and the cohesive attraction between the two molecules, predominates. Now suppose the repulsive intensity of the similar poles, E, E, and e, e, to be so much increased, as to extend beyond the semi-diameters of the molecules. Such an increase of

repulsive intensity may be imagined to cause
the molecules to turn round on their common
points of cohesion, M, m, till the chemical axes,
$E\ e$, and $E\ e$, become at right angles to each
other; as is represented in Fig. 13; or rather as

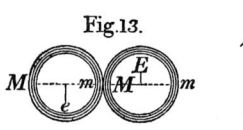

Fig.13.   Fig.14.

in Fig. 14. In Fig. 14,
a view is supposed
through the two mole-
cules, in the direction
of the conjoined diameters of cohesion, $M\ m$,
and $M\ m$, of Fig. 13; in which view, of course,
$E\ e$, and $E\ e$, are supposed to represent the chemical axes of the two molecules, at right angles
to each other. We may therefore conceive, that
in the liquid state of bodies, the chemical axes
of their adjoining molecules are at right angles
to each other; or in some position, intermediate
to right angles and that of parallelism. When
the chemical axes of adjoining molecules are
precisely at right angles, the chemical polarities,
and the cohesive attractions, are both exactly
balanced and neutralized; so that the points
M, and m, have neither a tendency to unite, nor
to separate; but remain simply in contact.
Hence the molecules of such a body will be
all disassociated, and free to move among one
another; and if we suppose each molecule, at
the same time, to be surrounded by its atmosphere of caloric, so thin as not altogether to

remove mechanically the molecules beyond one another's influence ; we have probably as clear an idea of the molecular constitution of a liquid, as we are capable of forming.

Section V.

Of the Third, or Gaseous Form of Bodies.

WE proceed to examine the most perfect form of fluidity, that of gas ; and shall, in the first place, consider the molecular arrangement of bodies existing in the gaseous form ; which will enable us to explain further, the subject of latent heat. Still illustrating our views, by the changes produced in water by increase of temperature ; we shall adopt *steam*, as an example of a gas. Let us suppose the same constant flow of heat to be entering into a portion of water, which we before supposed to be entering into the ice : the water continues to increase in temperature, in volume, and in capacity for heat, till it arrives at the boiling point ; at that moment, the temperature ceases to be augmented, however much we may urge the application of heat ; and the water is converted into a transparent gas, well known by the name of steam. For the conversion of water

into steam, however, under the ordinary circumstances of atmospheric pressure, it has been
found that nearly 1000 degrees of heat are necessary; which large quantity of heat actually
becomes latent or disappears; since the temperature of the steam formed, never exceeds 212°,
that of the water at the boiling point. What
becomes of these 1000 degrees of heat? We
may conceive one portion of the heat to become
latent, in the first of the two ways described
above; that is to say, the water in the act of
being converted into vapour, is much augmented
in volume; and into this augmented volume, as
into a sort of vacuum, a portion of the heat may
be supposed to rush, and to become insensible;
but another portion of heat obviously goes to
augment the molecular polarities of the water;
which, in the case of steam (and in all gases),
may be imagined to be arranged in some such
way as the following :—

Fig. 15 gives a view of two molecules of water,

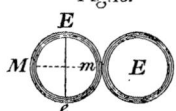

in which the chemical axes, $E\,e$,
and $E\,e$, are at right angles to
each other; the same position
as in Fig. 13, and Fig. 14, but
otherwise represented. In this position of the
chemical axes, as before stated, not only the
chemical polarities, but also the cohesive attractions, are exactly balanced and neutralized.
The temperature, therefore, which communicates

to the polarizing energies of the chemical axes
of the molecules of any body, such intensity,
that these axes are brought into a position, in
which they are at right angles to each other,
may be considered to be the temperature, at
which that body attains perfect liquidity. The
next augmentation of heat acquired by the
molecules of any body, may be conceived to
extend still farther the polarizing energies of

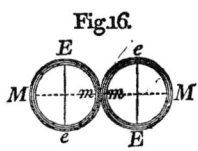

Fig.16.

their chemical axes; so as to
cause the increased intensity of
their polarities, to again determine the chemical axes into a
state of parallelism, as in Fig. 16.
In this state of parallelism, however, the position
of the chemical axes is just the reverse of that
which they assume in the parallelism of solidity,
Fig. 12 ; the similar chemical poles being, not in
the same, but in *opposite* directions. When the
chemical axes of the molecules of any body are
brought into this reverse state of parallelism, the
natural repulsion of their similar chemical poles
may be conceived to have reached its maximum.
The points of cohesion, *M, M,* and *m, m,* will
be also reversed ; that is to say, the similar, and
mutually repellant points, *m, m,* will be approxi-
mated. Molecules, whose chemical axes are
thus situated, therefore, instead of cohering, will
have a tendency to recede from each other ;
that is to say, will become *self-repulsive,* and

free to move in any mode or direction, their energies or circumstances may require.

Hence, as there are two conditions of the solid aggregation of the molecules of matter; namely, aggregation depending on the cohesion of the molecules of the same kind of matter; and aggregation depending on the union of the molecules of different kinds of matter: so there must be likewise two modes of repulsion ; namely, homogeneous repulsion, or the mutual repulsion of the molecules of the same kind of matter, which is opposed to cohesion, and from whence chiefly arises the gaseous condition of bodies; and heterogeneous repulsion, or the mutual repulsion of molecules chemically re-pellent, which prevents such molecules from uniting together chemically.

But here a question arises: What is the molecular arrangement of bodies in the state of *vapour ;* the state which water is liable to assume at all temperatures below 212°; for instance, at 32°? According to an hypothesis, to be presently mentioned, a given volume of steam, at 212°, the boiling temperature of water, contains the same number of self-repulsive molecules, as a similar volume of air under the same temperature and pressure, and therefore has the same elasticity. But the elasticity of the vapour of water at its freezing temperature, 32°, is only

equal to about one-fifth of an inch of mercury ;*
hence the same given volume of aqueous vapour
at 32°, will only weigh about 1-150th of what
aqueous gas, or steam, ought to weigh, supposing
that water could exist as a permanent gas at 32°,
and under a pressure of thirty inches of mercury.
The molecules of the vapour of water, consequently, will be five or six times further apart,
than those of perfectly gaseous bodies, at the
same temperature, and under an equal pressure.†

* The elastic force of vapours increases with their temperatures,—a phenomenon that may be represented, either by the
greater or less angle imagined to exist between the axes of contiguous molecules; or by the greater or less velocity of these
molecules on their axes; on which greater or less velocity, indeed, the angle formed by the axes of contiguous molecules
may be supposed to depend. See Appendix.

† Supposing it were possible for steam to exist at 32°, of
course at this temperature, its weight would bear to that of air,
the same proportion it bears at 212°; the proportion, namely, of
5 to 8. One hundred cubic inches of steam, at 32°, ought, therefore, to weigh 20·49375 grains; or 5-8ths of 32·79 grains, which
is the weight of 100 cubic inches of air at 32°. But the weight
of 100 cubic inches of steam at 32° is only ·1366 grain, or
1-150th that of air. The number of molecules in steam at 32° is
consequently only 1-150th of those in air at 32°. Hence this
diminution of the number of molecules of aqueous vapour, if we
suppose them to be diffused equally throughout the same space of
100 cubic inches, must of course, as stated in the text, cause
them to be between five and six times further apart, than those
of air, or of any gas at the same temperature.

Let us now close these illustrations of the molecular arrangement of matter, by enquiring, how far the above suppositions respecting gaseous bodies, accord with their essential properties; viz., with their self-repulsive or diffusive properties; their equable expansion by heat; their increase in volume in the inverse proportion of the force with which they are compressed; and with their similar capacities for heat.

Of the diffusion of Gaseous Bodies.—For the facts connected with this most important subject, we are principally indebted to Dr. Dalton and to Mr. Graham; the latter of whom has shown, that when any gas is confined in a vessel furnished with a very narrow aperture, or with a porous plug, an interchange between the confined gas and the external air, immediately begins to take place, through the communicating aperture; and that this interchange continues to go on to a certain point, which, with respect to the same gas, appears to be uniform; but differs in different gases, according to a certain law depending upon the specific gravities of the gases. Different gases, also, whether of the same, or of different specific gravities, and however they may be introduced into the same vessel, speedily become mixed uniformly throughout. These facts evidently indicate a species of self-repulsive influence among the molecules of the *same* gas, which appears to be satisfactorily accounted for by our hypothesis. The argument is very simple and obvious : Two molecules of the *same* matter have a tendency to cohere and to form a solid, when the chemical polarities of these molecules are similarly arranged, and do not extend beyond the semi-diameters of the molecules ; but two molecules of *different* matter, under circumstances precisely alike, remain passive, and have no tendency to cohere. Hence, while two molecules of the same matter, having the intensity of their polarities much increased, and their chemical poles, consequently, reversed, repel each other, or become *self*-repulsive ; two molecules of different matter, still retaining their mutual passiveness, do *not* repel each other.

There is reason to believe, that the phenomena of diffusion are not confined to bodies perfectly gaseous : but exist also in the imperfectly gaseous state of bodies, termed *vapour ;* of which the vapour of water may be considered as the most familiar example. Phenomena very similar to, if not identical with, the above, also exist in liquids, and perhaps in solids. Thus the molecules of certain matters diffused through a liquid, as through water, may, in some cases, be supposed to exert a self-repulsive influence on each other ; by which supposition only, does their equal diffusion through a large mass of liquid seem explicable. Even in the solid state, as above observed, something of the kind appears to exist, especially among organized bodies; which apparently owe several of their most remarkable properties to the diffusion of active self-repulsive molecules throughout their substance.*

Of the equal Expansion of Gaseous Bodies by Heat.—With respect to the second essential property of gaseous bodies, that, under the same temperature and pressure they all undergo equal expansion by an equal increase of heat; this seems to be explicable only on the supposition that—*all gaseous bodies, under the same pressure and temperature, contain equal numbers of self-repulsive molecules:* a most important conclusion, as we shall see hereafter, and one, which at present we are anxious a little further to illustrate. Admitting the fact to be, as it is, undeniable, that within the ordinary limits of experiment, all perfectly gaseous bodies expand equally by similar increments of heat; *if* different gases contain *unequal* numbers of self-repulsive molecules, those gases which contain the least number of molecules, must exert the greatest power, and consequently have the greatest disposition to expand; in other words, the expansive energy of the molecules of a gas, must increase as their number diminishes; and

---

* For further observations on the diffusion of gaseous bodies and of vapours, and their operation in the economy of nature; see the Appendix.

not only so, but in order to produce the effect stated, the expansive energy must increase, neither more nor less, but exactly, as the number diminishes—a law which when applied to extreme cases, becomes obviously absurd. Further it may be observed, in corroboration of the hypothesis advanced, that in the gaseous state, the molecules of bodies may be considered as having undergone the utmost effects, that any increase of heat can produce upon them. All their interstitial vacuities may be supposed to be already saturated with it; while an atmosphere may be supposed to surround each molecule, keeping them individually at a considerable distance from each other : their polarities also may be supposed to have undergone their ultimate change; so that no more heat can be rendered latent by inducing further changes, except in degree; which degree may be supposed to be common to all gases.   Hence every molecule of matter, in the gaseous state, when subjected to similar pressure and temperature, may, without reference to its other properties, be supposed to be in circumstances exactly similar; and consequently liable to be affected in an exactly similar manner, by all further increments of heat.

Of the inverse Relation of the Volume to the Compressive Force.—Nearly the same remarks apply to this law as to the preceding; for were the numbers of molecules in each gas supposed to be *unequal,* the diminution of volume under similar pressure ought to vary also, which is not the case, at least in the more perfect gaseous bodies ; neither this observation, or those in the former paragraph, apply to vapours.

Of the similar Capacity of Gaseous Bodies for Heat.—The best experiments seem to show, that under equal pressures, the same volumes of all gases have the same capacity for heat—a circumstance quite according with the other phenomena. Hence, for the reasons assigned in the two foregoing paragraphs, and for other reasons which might be mentioned, we have been induced to adopt the hypothesis already stated, that, *under equal pressures, and at the same temperature, all bodies in a perfectly gaseous state, contain equal numbers of self-repulsive molecules.**

---

* It is proper to observe, that these views were adopted by the author, long before he was aware of the existence of the essays on the subject, by Messrs. Avogadro, Ampere, and Dumas. Indeed he was unacquainted with those of Dumas, which most nearly resemble his own, till he saw them alluded to in Mr. Johnston's recent report on chemistry, in the Transactions of the British Association. Mr. Donovan seems to consider the above hypothesis as untenable ; but we think his arguments entirely inconclusive. See Giornale di Fisica, sec. ii. tom. viii. p. 1 ; Annales de Chimie, tom. xc. p. 43 ; a Treatise on Chemistry, by Mr. Donovan, in Lardner's Cabinet Cyclopædia, p. 379 ; and the Introduction to Dumas's Traité de Chimie appliquée aux Arts ; which excellent work· the author had been prevented from perusing, by the nature of the title.

SECTION VI.

Other Properties of Heat. Of Heat in Motion.

HEAT appears to be in a constant state of motion and of interchange between different bodies, among which it finally settles into a state of equilibrium. If accumulated in any body, this accumulation cannot be preserved ; but the excess will fly off, in spite of all we can do to the contrary, and sooner or later, the equilibrium will be restored. This motion of heat apparently takes place in three ways, which a common fire-place very well illustrates. If, for instance, we place a thermometer directly before a fire, it soon begins to rise, indicating an increase of temperature. In this case, the heat has made its way through the space between the fire and the thermometer, by the process termed *radiation*. If we place a second thermometer in contact with any part of the grate, and away from the direct influence of the fire, we shall find that this thermometer also denotes an increase of temperature ; but here the heat must have travelled through the metal of the grate, by what is termed *conduction*. Lastly, a third thermometer placed in the chimney, away from the direct influence of the fire, will also indicate a considerable increase of temperature; in this case, a portion of the air, passing through and near the fire, has become heated, and has *carried* up the chimney the temperature acquired from the fire. There is at present, no single term in our language, employed to denote this third mode of the propagation of heat; but we venture to propose for that purpose, the term *convection*,* which not only expresses the leading fact, but also accords very well with the two other terms. Each of these modes of the propagation of heat possesses certain peculiarities, on which we proceed to make a few remarks.

Radiation of Heat.—Heat radiates *in vacuo* in all directions equally, and with immeasurable velocity. Heat radiates also through all gaseous bodies, and more or less, through transparent media. Radiation goes on at all temperatures; but the quantity of heat radiated in a given time, bears some proportion to the excess of the temperature of the radiating body, above that of the surrounding medium. Radiant heat is

---

* *Convectio*, a carrying or conveying. We state these three modes of the propagation of heat in accordance with popular language; and as expressive of the modes in which heat is *apparently* communicated. The convection of heat, philosophically considered, is in reality a modification of the conduction of heat; while the conduction of heat may be viewed as an extreme case only, of the radiation of heat.

capable of being *reflected* like light, (to be presently noticed,) and, indeed, obeys altogether somewhat similar laws. Those surfaces, however, that reflect light most perfectly, are not equally adapted to the reflection of heat. Metals in general, and particularly when highly polished, are the best reflectors of heat; while glass, which reflects light most perfectly, reflects comparatively little heat ; thus tin-plate reflects about eight times as much heat as a glass mirror. The radiation of heat is much influenced by the nature and state of the surfaces of bodies. Thus a surface coated with lamp-black, radiates eight or nine times as much heat, as a polished surface of tin or silver ; and in general, polished surfaces, particularly of metal, radiate much less than other surfaces. As might be expected, this difference of radiating power exerts great influence in the cooling of bodies ; thus warm water retains its heat much longer in a bright tin vessel, than in the same vessel coated with linen, paint, or particularly lamp-black. Radiant heat is *absorbed* with different facilities by different surfaces. The absorbing power of surfaces seems, indeed, to vary directly as their radiating power; and inversely as their reflecting power. That is to say, surfaces receive heat by radiation, nearly with the same degree of facility as they give it off ; while those that reflect most, of course, must absorb least ; a surface covered with lamp-black, for example, receives in a given time, eight or nine times as much heat by radiation, as a polished tin surface receives. From these remarks, it will be readily inferred, that the *colours* of bodies may have considerable influence in the radiation and absorption of heat; now such is found to be the case; and the darker the colour of a body, the more readily it gives off, and absorbs, radiant heat. Radiant heat has the power of *passing through* transparent bodies, as glass. This power, however, varies according to the thickness of the glass, its relative position to the radiating body, and a variety of other circumstances, not well understood; but generally speaking, heat of great intensity, and particularly solar heat, as before observed, obeys laws, more or less analogous to those of light under similar circumstances. Heat of low intensity, on the contrary, as that from boiling water, is said to present some peculiarities in its motions. These peculiarities, however, are such as scarcely to require to be here detailed.

Conduction of Heat.—The conduction of heat is chiefly confined to solid bodies; and as solids exist of every degree of consistency and density, from perfect fluidity up to perfect hardness, the conducting power varies in like manner. Hence the laws of conduction, and those of radiation, have a mutual dependance; and, in fact, the laws of conduction, may be considered as only extreme cases of the laws of radiation. The conduction of heat through bodies seems to take place equally in all directions. In general the densest bodies, as metals, stones, hard woods, &c., have the greatest conducting power; though these differ exceedingly among one another. Porous bodies in general are bad conductors; and of such bodies, charcoal may be considered as one of the worst conductors. Among substances employed as articles of dress, hare's fur, and eider down, are the worst conductors, and flax, the best. The relative conducting powers of substances of this class seem to depend much, on the quantity of air enclosed within their interstices; and on the power of attraction by which this air is retained or confined. The conducting power of liquids and of gases is very limited; though under certain circumstances, they *appear* to possess this power in a high degree. But this power is only *apparent;* and heat is principally communicated through liquids, and also through gases, by the third process above alluded to, viz. *convection.* By convection however, heat is chiefly propagated in one direction, that is to say, *upwards;* hence almost any degree of heat, may, for a long time, be applied to the upper surface of a liquid, or of a gas, without materially affecting the temperature of the lower surface.

P.   G

Such are the principal phenomena connected with the motion of heat; but before we proceed to speak of the *sources* of this wonderful agent, we have yet to consider another imponderable principle of the utmost importance, and intimately connected with heat; viz. *Light*.

Section VII.

Of Light.

The laws of the motion of light, of its reflection, refraction, polarization, &c., properly belong to another department; we shall, therefore, only briefly describe them here, and endeavour to point out the general connexion and analogy they bear to the phenomena of chemistry; and more especially, to the phenomena of heat and of electricity.

Radiation, or Motion, of Light.—Light radiates or moves in straight lines with such inconceivable velocity, that it occupies only about eight minutes in travelling from the sun to our earth; so that it must move at the rate of nearly 200,000 miles in a second! At the same rate it would occupy about four hours to travel to us from the planet Uranus, the present *ultima Thule* of our system; hence if this planet were at any given instant suddenly annihilated, we should not miss it for four hours afterwards; and when we look at it, we do not see it where it actually is at this instant, but where it was four hours previously. A cannon ball, when first shot from the cannon, moves with a velocity of between 2000 and 3000 feet per second; supposing, therefore, it could retain its initial velocity, it would scarcely move in a year, as much as light moves in a single second! The utmost velocities of the earth and other planets, in their orbits, or on their axes, scarcely exceed 30 or 40 miles in a second. Hence the utmost velocity that we are acquainted with as possessed by ordinary matter, and therefore, the utmost perhaps, of which such matter is capable, only amounts to the 1-5000th or 1-6000th of that of light! These striking facts are mentioned with the view of conveying some notion of the immensity of space; and of the wonderful velocity with which it is, in every direction, penetrated by light. They seem also to show, that if light be material, the matter of which it is composed must exist in a state of tenuity, totally different from the ponderable matter we are acquainted with, which actually seems incapable of such velocity.*

If we consider heat and light to consist of polarized molecules in the self-repulsive state, and to obey the same laws that ponderable matters in the gaseous state obey, which is exceedingly probable; the radiation of these im-

* Pouillet, Elémens de Physique et de Métérologie, tom. III. p. 216.

ponderable bodies will be analogous to the diffusion of gaseous bodies; and by knowing their
velocity and applying the same law, we may
deduce their comparative gravities.

Reflection, and Refraction, of Light.—In free
space, as before observed, light moves in straight
lines; but when a ray, *R*, Fig. 17, falls upon a

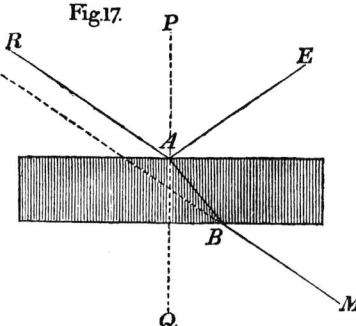

polished surface, as
of glass, a portion of
it is reflected in the
direction *A E*, and
the angle, *R A P*,
called the angle of
incidence, is always
equal to the angle,
P A E, called the
angle of *reflection*. Another portion, *A B*,
passes through the glass, but instead of continuing to move in the same straight line, is bent
considerably out of that direction, towards the
perpendicular *P Q;* it then makes its exit at *B*,
and goes on in the direction *B M*, parallel to its
original direction, *R A*. This portion of the
ray is said to have undergone *refraction;* a term
indicating that its natural course has been
broken. Such are the general facts; and the
study of their laws, varieties, and peculiarities,
as modified by different media, constitutes the
science of optics; a branch of knowledge not
falling within our present enquiry. In connection with this part of our subject, it only

remains to observe, that in passing through the most transparent bodies, much light is lost, by absorption and in other ways. So also when light falls upon metallic bodies, such as polished silver, about one-half only is reflected, while the other half is absorbed and lost. Different substances, however, differ materially in these respects: thus from the experiments of M. Bouguer and M. Lambert, it appears, that in fluids, transparent solids, and metals, the quantity of light reflected, *increases* with the angle of incidence, reckoned from the perpendicular ; whereas in white opaque bodies, the quantity of light reflected, *decreases* with the angle of incidence.* We shall hereafter, have occasion to revert to these curious facts.

Polarization of Light.—The next property we have to notice, is what is called the *polarization* of light. Let us suppose Fig. 18,

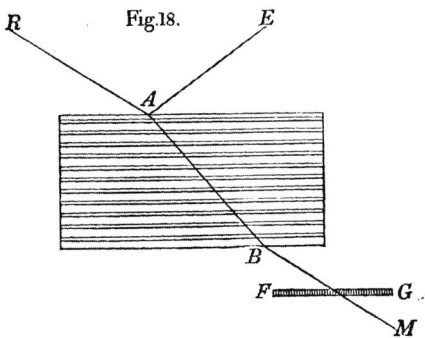

to represent a bundle of plates of thin window-glass, bound together in the manner indicated. Let *R A* be a ray of light falling on the

* See article Optics, p. 67, and 68, in the Library of Useful Knowledge. Where the original observations are to be found, which are there referred to, we do not at present know.

upper plate, at an angle of incidence of about
56°; a portion of the ray will be reflected, and
will move in the direction *A E;* while another
portion of the ray, *A B*, will pass through the
bundle of glass plates onwards to *M*, according
to the laws of reflection and refraction already
stated.   Now these two rays *A E*, and *B M*,
possess remarkable properties, similar to one
another in most respects, but directly opposed in
another. Of these properties we shall endeavour
to give a general idea.

If the ray of light *R A*, after falling upon the
vertical glass *A*, Fig. 19, at an angle of incidence
of 56°, be received on a plate of glass, *C*, placed

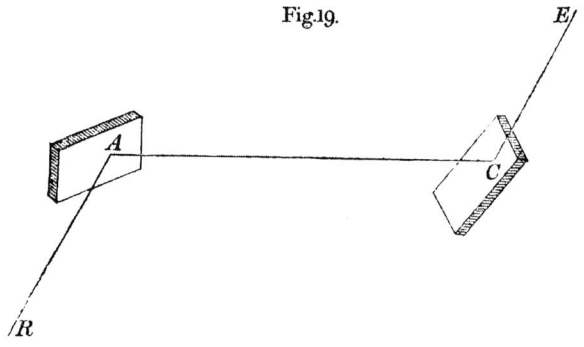

Fig.19.

at the same angle of incidence, and be then
reflected from *C* to *E;* in the position intended
to be shown in the figure, when the ray *R* is·
first reflected in a horizontal plane, *R A C*, and
then in a vertical plane, *A C E*, the ray *C E*
becomes so weak as to be scarcely visible, the

whole of it having passed through the glass *C*. But if the glass *C* be turned round 90°, (the ray *A C* being supposed to be the axis of motion,) so that the ray *C E* be reflected horizontally; instead of passing through the glass *C*, as before, the whole of the ray *C E* will be reflected. If we continue to turn the plate *C* upon the axis *A C*, round the entire circle, these alternations of transmission and reflection, will be found to take place in the same manner, at the two other quadrants 180°, and 270°. Hence the ray *R A*, by reflection, has acquired properties altogether new; it is said in short, to have acquired *polarity*, or to have become *polarized*. Now recurring to Fig. 18, the ray *R A*, in that figure, will of course follow the same laws as the ray *R A*, in Fig. 19; that is to say, the ray *A E* will have acquired polarity by reflection. Let us now consider what has happened to the refracted ray *B M*, in the same Fig. 18. This ray *B M* will also be found to be polarized; but if we receive it on a glass plate, *F G*, at the polarizing angle of 56°, we shall find that it will refuse to be reflected; whereas the reflected ray *A E*, does not refuse to be again reflected, unless the plate *F G* be turned round 90°; or into a plane at right angles to that plane in which the refracted ray *B M*, had refused to be reflected. Hence we conclude, that when a ray of light is incident at the polarizing angle, upon a transparent body, the whole of the reflected light is polarized ; while the whole of the transmitted light is also polarized ; but *in a plane at right angles*, to that in which the reflected ray is polarized.

Such is the general law ; and it may not be amiss to allude briefly to another familiar illustration of it. Every one is acquainted with the mineral called Iceland spar, and with the singular property which this mineral possesses of forming a double image of objects seen through it, or its property of double refraction ; in other words, when a ray of light falls on a crystal of such spar in a particular direction, the ray is separated into two. Now it is a remarkable fact that if these two rays be examined in the way before directed, when speaking of reflected and transmitted light ; it will be found that both are polarized, but that the two rays are polarized in planes *at right angles to each other:* that is to say, the ordinary transmitted ray is polarized like the ordinary ray, transmitted through the bundle of glass plates ; while the extraordinary transmitted ray is polarized like the ray reflected from these plates. Many bodies are similarly constituted ; while others have two or more planes or axes of double refraction, giving origin to a variety of curious and beautiful properties ; which it would be quite foreign to our present purpose to detail further.

Decomposition of Light.—When a ray of light, *R*, Fig. 20, traverses a prism, *C D F*, instead of passing onward in the direction *Y*, it is refracted into the spectrum *E e;* which spectrum when received upon the screen, *A B*, will be found to consist of seven different colours, in the order, and of the kind described, each having, of course, different refractive powers; the red being the least, and the violet the most, refracted from the original direction *R Y*, of the

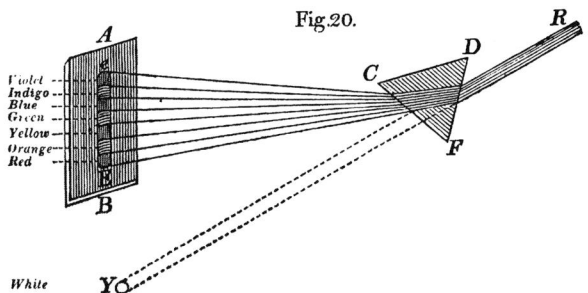

Fig. 20.

solar beam. This oblong image is called the *solar*, or sometimes, the *prismatic spectrum;* and Sir Isaac Newton found that each colour consists of light no longer separable, like white light, into others, but having uniform refractive properties: hence he called all the seven colours simple, or homogeneous; in opposition to white light, which he called compound, or heterogeneous.* This important fact presents a clue to,

---

* Sir David Brewster has lately shown that there are, in fact, but three simple colours, the *red*, the *yellow*, and the *blue ;* and

and exhibits the general law, which regulates the endless variety and change of colours; as bodies appear to have this or that colour, according as they have the power of reflecting or transmitting the rays of this or that colour, and of absorbing or reflecting the rest; while white bodies reflect all, and black absorb all. Besides colour, it has been likewise noticed, that different portions of the prismatic spectrum possess different *heating*, and *chemical*, or *electrical*, properties. These properties vary in some respects, according to the nature of the prism employed. In general, the heating power increases towards the red ray: while the chemical power seems to be regulated in some degree by the nature of the colour; but is greater, though of opposite character, at the two extremities, than in the centre of the spectrum, where it appears to be nearly null. The chemical properties of light, however, are by no means well understood, and have not received the attention which they merit.

Upon an attentive consideration of the phenomena of heat and of light, and a careful comparison of them with the general phenomena of polarizing forces; it is impossible not to be

that each of these colours exists throughout the spectrum. Hence, like the different electric and magnetic energies, these elementary colours, or, at least, the *red* and the *blue*, (the *yellow* probably being merely resultant,) appear to be incapable of entire separation from each other.

struck with the close analogy that prevails throughout the whole. The phenomena of heat have hitherto been very imperfectly studied, and in consequence we are much less able to trace the analogy ; but the phenomena of light, from their obvious and striking characters, have attracted more attention, and are much better understood. To enter further upon the enquiry here would be quite foreign to the object of this treatise ; we cannot, however, in concluding these remarks, refrain from repeating an opinion already expressed, that the molecules, both of heat and of light, possess polarities precisely similar to those of ponderable bodies ; and that not only the chemical agencies of these principles, but those phenomena of light also, at present so beautifully illustrated by the hypothesis of undulæ, will be hereafter found to admit of explanation, on the more probable supposition of molecular polarity.*

* In the Newtonian hypothesis of *fits of easy transmission and of easy reflection*, the molecules of light may be regarded as little magnets revolving rapidly round their centres while they advance in their course, and thus presenting alternately, their attractive and repulsive poles. (See Discourse on the Study of Natural Philosophy, by Sir J. Herschel, p. 253.) In our hypothesis, the chemical axes of all self-repulsive molecules are supposed to be reversed, by which reverse arrangement the poles of contiguous molecules are rendered *dissimilar* and attractive ; while by the same arrangement, the cohesive equatorial energies of contiguous molecules are necessarily *similar* and repulsive. (See

SECTION VIII.

Of the Sources of Heat and Light.

THE principal and obvious sources of heat and
light are the sun, electricity, mechanical action;
change of physical condition, change of chemical
condition; and organic action.

Fig. 16, p. 69.) Now the unfavourable position, and greater
distance, of the chemical polarities under these circumstances,
may be supposed to so much diminish the intensity of their attractive powers, that although from their perfect equality in opposite directions, they are able to preserve the chemical axes in
a state of parallelism; these chemical polarities are not able to
overcome the superior influence of the equatorial repulsion between contiguous molecules, and thus to prevent the recession
of these molecules from each other. Hence, when a series of
self-repulsive molecules move onwards in virtue of their self-repulsive powers, (as in the radiation of light, &c.) the equatorial,
or cohesive diameters, of the molecules, will always be in the
line of motion, and each successive molecule will present alternately an opposite polarity; while the chemical axes, of course,
will be all in the same plane, and transverse to the line of motion.
Such will be the order in a single series of molecules in motion;
but when a number of series move onward together, as in common solar light, there is reason to conclude, that the molecules
of contiguous series have a tendency to arrange themselves
thus ·.·.·.·.· with the chemical axes at right angles to each
other. Those who are interested in the subject of light, will
perhaps readily conceive how such arrangements may be applied,
to explain the various phenomena we have been considering.

The sun is the most obvious and unvarying source, from which both heat and light are communicated to our earth. The nature of the sun, however, and the mode in which that wonderful supply of heat and light is maintained, are quite unknown to us, and will probably always remain so. Electricity is another source of heat and light, which are developed at the moment of the equilibrium of the two energies; and some of the most intense degrees of heat and light that have been produced, have sprung from a galvanic apparatus. The sudden condensation of air is likewise a source from which heat and light are often both extricated; on principles, that it will not perhaps be difficult to understand, from what has been stated. The extrication of heat by percussion and condensation appears to be limited, but its extrication by friction seems to be boundless; that is to say, so long as friction is kept up, will heat continue to be extricated; but whence the heat is derived, does not appear to be capable of satisfactory explanation; unless we suppose a perpetual decomposition and recomposition to take place, which is not improbable. Another fertile source from which heat is derived, is the physical change of condition which bodies are constantly undergoing in nature; such as the conversion of gases into liquids; of liquids into solids; &c. by taking advantage of which conversions, we can accumulate heat at will; as for instance, by the condensation of steam. When these physical changes however are associated with chemical changes, as is very often the case, the most striking effects are produced. Of this kind are all the phenomena of combustion; the most common source of artificial heat: which phenomena consist of nothing more, than the rapid chemical union of certain bodies with others; and generally, with the principle termed *oxygen*. Nearly allied to chemical action, and perhaps identical with it, is the extrication of heat by organic changes, or what is termed *animal* heat; a subject we shall have to consider in a future part of this volume.

In concluding, for the present, our remarks on heat and light, it only remains to observe, that the phenomena, and laws of motion, of these subordinate agents, are all of the utmost importance; as constituting limitations and principles of action, to which the Great Author of nature most rigidly adheres in his operations. Hence, whether we view the distribution of heat and of light on the large scale, as regulating climate; or whether we view them with reference to the most trifling particular, as the clothing of a bud or of an insect; we find the same beautiful adaptation and contrivance, every where exemplified, to ensure, or to evade, the agency of these all-important principles. The wonderful arrangements connected with heat and light, will however fall more naturally to be considered hereafter; we shall therefore, defer what we have to say further on these subjects, till we come to speak of meteorology.

Section IX.

Recapitulation, and General Observations on the Subjects treated of in the preceding Chapters.

In the preceding observations, we have endeavoured to give a connected sketch, of the nature and operation of molecular forces: perhaps it will still further facilitate the understanding of the subject, if we recapitulate briefly the leading facts; so as to point out to those who may not be inclined to peruse the foregoing details, the analogy that prevails throughout the whole.

1. In the first place, we attempted to show, that the forces which determine molecular union, can scarcely be those of mere gravitation, in their ordinary forms at least;. but that some other modification of force is necessary to account for the phenomena.

2. By assuming the molecules of bodies to be virtually spheroidal, and endowed with two kinds of polarizing forces, the one operating *axially,* and the other *equatorially,* we attempted

to show how the phenomena of simple crystallization might be explained; and we corroborated
our argument, by demonstrating that the electric
and magnetic forces are actually related to each
other, precisely as we assumed the energies of
our molecules to be.   Hence we ventured to
draw the conclusion, that electricity and magnetism, if not identical with, at least represent,
or are analogous to those forces, the existence of
which among ponderable bodies, we assumed as
necessary to account for the phenomena of crystallization.   Further, we attempted to render it
probable, that the molecules of the imponderable
principles, heat and light, possess polarities precisely analogous to those of ponderable bodies;
and that many of their peculiar phenomena
depend upon these polarities.

3. In attempting to account for the different
forms assumed by bodies; we supposed that in
the *solid* form, the molecules are so arranged as
to attract each other, according to certain laws;
that in the *liquid* form, they are so arranged as
neither to attract nor repel each other; and that
in the *gaseous* form, the arrangement of the energies of the molecules is such, as to render them
mutually repulsive.   Moreover, by assuming that
those molecules which possess the property of
attracting each other in the solid form, in preference to others; retain a similar relation in
the gaseous form, and repel each other in pre-

ference to others; we attempted to account for many of the well known phenomena of gaseous bodies.

4. Lastly, we attempted to show, that the phenomena of *radiation* among the molecules of imponderable bodies, are precisely analogous to the phenomena of diffusion and mixture among the molecules of ponderable bodies, when in the liquid and gaseous states; and that consequently, the same laws are strictly applicable to both.

With respect to the reasons which have induced us thus to enter into the dry details of molecular action; and which may seem to require some apology to our readers, they are chiefly twofold: In the first place, as connected with the particular business of the present treatise, it has been our object to convey to the general reader, some idea of the wonderful operations which are constantly taking place in every particle of matter which he sees around him; or, to use the language of Paley, some notion of the " concealed and internal operations of the machine." These operations may not be as we have represented them.; they may in fact be altogether different: but be this as it will, a perusal, however cursory, of what has been stated, can hardly fail to accomplish one purpose we had in view; viz. to show to the most incurious and superficial reader, that in the mi-

P.   H

nutest fragment of matter; and in the commonest and simplest operations of nature; which he is altogether too apt to overlook; the most wonderful and extraordinary arrangements, *must* take place: arrangements which, if duly considered, are calculated fully as much, if not more, than those connected with the more obvious phenomena, to excite his wonder; and, at the same time, to display the omnipotence of the great Creator. The second object we aimed at, was, as just remarked, to give a connected sketch of molecular forces; and, by placing the different operations in a point of view, in which we believe they have not hitherto been considered, to display the striking analogy that prevail throughout the whole.

Finally, it remains, before we close, to state briefly the arguments deducible from the divisibility and molecular constitution of matter, with reference to our present subject. These arguments may be considered under the three following heads: first, that matter has not *always* existed in its present form: secondly, that it could not have existed in its present form by *chance:* thirdly, and consequently, that it must have been the work of a voluntary, and *intelligent* Being. Other deductions might doubtless be made from what has been stated; but these we purposely avoid, and confine our arguments, as much as possible, to grounds admitting of no controversy.

In the first place, the divisibility and molecular constitution of matter, seem to prove beyond a doubt, that it cannot have *eternally* existed in its present state.

Although we can form no idea of what matter would be, without its molecular properties; there is yet nothing in these properties which can induce us to believe, that they are *necessary* to the mere *existence* of matter. On the contrary, we have seen, that matter possesses qualities, (those of gravitation), of a more primordial kind ; to which its molecular properties are apparently secondary or subordinate. But if these subordinate properties be not necessary to the existence of matter, matter might *possibly*, at some time, have existed without them. Now this very possibility seems incompatible with eternal existence ; for what *can* happen, *may* have happened, at some period : the eternal (passive) existence of matter, therefore, ought to involve incapability of change. Hence, as the molecular constitution of matter, cannot be shown to be *necessary* to the existence of matter ; that molecular constitution cannot be *proved* to be eternal : moreover the difficulty of such a supposition is exceedingly increased, when we consider the characteristic property of matter in the molecular state ; viz. *the endless repetition of exactly similar parts.* It is to be observed also, that the above remarks apply to the supposition of only *one* form of matter ; but we shall see hereafter, that chemists recognise upwards of *fifty* forms of matter, all of an elementary character; at least, we cannot at present say, that one of these forms, is more elementary than another. Again, the number of molecules in each of these elementary principles, great as it is, is *limited*; the properties of the molecules also are fixed and definite; all which circumstances throw further insurmountable difficulties in the way of the supposition, that the whole have existed, as they now exist, from eternity. For how has it happened, it may be asked, that the number and properties of the elements, or that the number of the molecules of which these elements consist, are just what the economy of nature requires; and are neither greater, nor less, nor different? How has it happened, that what is supposed to be infinite in some respects, should be finite and limited in those respects, in which we are actually able to trace them; nay, what is more, most luckily finite and limited, just where they appear to be required to be so? He who can satisfactorily answer these questions, may contend with some prospect of success for the eternity of matter, and of its properties, in their present form. In the mean time, we assert without fear of contradiction, that the molecular constitution of matter is decidedly artificial; or to use the words of a celebrated writer, that the molecules of matter have all " the essential cha-

racters of a manufactured article," * and consequently are not eternal.

Secondly. If the present molecular constitution of matter has not always existed, it must
have been produced at some time, by some cause
superior to itself. Now this cause must have
operated either accidentally and by *chance;* or
voluntarily and under the influence of a *will.*

With respect to the first of these alternatives,
viz. chance ; the *endless repetition of similar parts*
presented by the molecular constitution of matter, seems absolutely to preclude this supposition. Do we not consider it a subject of wonder
to see even two or three things by chance alike ;
as for example two or three human faces?
Should we not consider the man absolutely mad,
who would attribute the uniform, or manœuvres,
of a regiment of soldiers to chance? and can we
then resist the argument in the infinitely stronger
shape, in which it is here presented to us? Thus,
as the idea of chance seems too monstrous to be
entertained for a moment by any rational being,
we are driven irresistibly to the other conclusion ;
viz. that the cause or agent who formed the
molecular constitution of matter, was a *voluntary* agent, or *Being;* and moreover, that this
Being possessed a *power* commensurate with his
will.

* Sir J. Herschel on the Study of Natural Philosophy, p. 38.

Thirdly. The agent or Being who constructed the wonderful system we have been considering, must have been as *intelligent*, as he was powerful.

We infer intelligence in an agent from the fitness and adaptation to certain ends exemplified in his works. When, for instance, we see a machine admirably fitted for the office it performs, we infer that the maker of that machine must have possessed intelligence. Now, if we judge of the molecular constitution of matter by this rule; we shall find, that there is not only the most extraordinary fitness and adaptation to circumstances displayed in its arrangements, as far as we can understand them, but evidently, much further; that is to say, the maker of this system, must not only have possessed intelligence; but intelligence infinitely surpassing our own. Thus at the very beginning, the adoption of the molecular form of matter, may be considered as indicating intelligence of the highest kind; for of all the forms that can be conceived to exist, the molecular form of matter seems best adapted to the purposes of creation. Indeed, on what other supposition, than that of the subdivision of matter into *minute similar parts*, could all those endless operations, which we see constantly going on in the world, be imagined to take place? The next circumstance which claims our attention, is the nature of the properties, with which the molecules of matter are endowed. These properties are truly astonishing, and calculated in the highest degree, to impress us with exalted notions, of the wisdom and power of their Contriver. Thus, what can be more wonderful, than that the self same *chemical* forces, differently directed, should produce, not only all that endless change of quality, and of condition, which we see around us ; and which are so beneficial, or even necessary, to our existence : but likewise some of the most terrible displays of energy in nature ; as for example, the utmost intensities of heat, of cold, and of light ; the terrors of the thunderbolt ; the irresistible demolition of the earthquake ! Nor, on the other hand, are the *cohesive* affinities existing among the molecules of matter much less wonderful, or important ; for if *similar* molecules had not been constituted with self-attractive and self-repulsive properties ; there would have been no aggregation of the same matter, into symmetrical groups ; no order or regularity ; no separation or purity ; in short, there would have been no common bond of union ; and the different molecules of the same matter would have been dispersed throughout nature, as accident, or other circumstances, might determine. Hence the present order of things could not have existed, unless the molecules of matter had been endowed with *both* these properties ; one of which, the *chemical,* as it were, goes before, and imperiously determines what molecules shall be combined or separated; while the other, the *cohesive,* silent and unobtrusive, follows in its train; and industriously assorting and arranging its predecessors' labours, here, perhaps, forms a diamond; or there, superintends the integrity of the atmosphere!

Such are molecular forces as they obviously appear to us; and such the arguments deducible from them. But when we attempt to go further, and inquire into the intimate nature of these forces; we not only find much, that is unknown to us; but much, that apparently surpasses our utmost conception! And what a still more sublime idea is this calculated to convey to us, of the wisdom and power of that Being, who contrived and made the whole! When, or where, do we naturally exclaim, did this Being exist? whence His wisdom? whence His power? There is, there can be, but one answer to these enquiries. The Being who contrived and made all these things, must have pre-existed from eternity—must have been omniscient—must have been omnipotent—MUST HAVE BEEN GOD!

CHAPTER IV.

OF CHEMICAL ELEMENTARY PRINCIPLES, AND OF THE LAWS OF THEIR COMBINATION.

In the preceding chapter we have endeavoured to show, that the minutest fragment of homogeneous matter cognizable by our senses, is composed of innumerable molecules; all of which are exactly alike in size, in shape, in properties, in short, of every kind; and we argued that these facts incontestibly prove, that the molecules of matter could not always have existed in their present form; nor have been formed by chance; but that they must have had a beginning; and have been the work of a Creator. Now when we consider the prodigious quantity of matter composing our globe, (to go no further,) or even composing a portion of it, as for instance, the mass of water existing in the ocean; and reflect that every individual molecule of this water possesses properties, exactly like those of the drop we formerly contemplated; our argument, already sufficiently convincing, actually overwhelms us with its force. Still however it admits of further corroboration; and we proceed now to show, that all this vast assemblage of molecules, so numerous, so diversified, so extraordinary as they are, may be reduced to a very few elementary groups ; upon the endless combinations and separations of which, all the phenomena of chemistry depend.

SECTION I.

Of Chemical Elementary Principles.

THE substances at present considered as elementary, amount to about fifty-four. Of these, several possess certain properties in common ; though they all differ from one another in subordinate particulars ; or in other words, are specifically different. Of the whole number, not above two or three exist, in any great quantity, in an uncombined state, at least in those parts of our globe to which we have access ; but the whole are wrapped up, as it were, and have their properties concealed, in compounds. Under ordinary circumstances, most elementary principles exist as solids ; but some of the more important occur in a gaseous form ; and one or two as fluids. A few of them are *apparently* of so little consequence in the world, that if they were annihilated, they would scarcely be missed ;

while others of them are so obviously necessary
to the existence of the present order of things ;
that the least derangement or alteration in their
proportion, or quantity, would be fatal to the
whole. Some of these elementary substances
exist in such enormous quantities, as to constitute a large proportion of the whole visible bulk
of our globe ; while others again, occur in such
minute proportion, at least within our reach, as
to be obtained with difficulty, and not without
elaborate research. With respect to the facility
with which they enter into combination ; and
the obstinacy with which they unite ; they differ
also, very remarkably ; a few of them combining
readily in a variety of proportions with almost
all the rest ; while some of the others, can be
scarcely made to combine under any circumstances. Lastly, the different effects, which different elementary substances are capable of exerting upon organic life, are equally striking. A
large majority of them indeed, may, in their
simple state, be considered of a deleterious
nature ; while three or four of them, on the
other hand, make organized beings what they
are ; and are necessary to their very existence.

Such are a few of the leading properties of
the elementary principles, as we are at present
acquainted with them. They have been arranged by Dr. Thomson under three great divisions, which he denominates, the *supporters of*

combustion; the *acidifiable bases;* and the *alkali-fiable bases.* The following table presents a summary of this arrangement.

TABLE.

I. *Supporters of Combustion.*

   1 Oxygen.
   2 Chlorine.
   3 Bromine.
   4 Iodine.
   5 Fluorine.

II. *Acidifiable Bases.*

   6 Hydrogen.
   7 Carbon.
   8 Azote.
   9 Boron.
  10 Silicon.
  11 Phosphorus.
  12 Sulfur.
  13 Selenium.
  14 Arsenic.
  15 Antimony.
  16 Tellurium.
  17 Chromium.
  18 Uranium.
  19 Vanadium.
  20 Molybdænum.
  21 Tungsten.
  22 Titanium.
  23 Columbium.

III. *Alkalifiable Bases.*

Alkaline Bases.
  24 Potassium.
  25 Sodium.
  26 Lithium.
  27 Calcium.
  28 Magnesium.
  29 Strontium.
  30 Baryum.

Earthy Bases.
  31 Aluminum.
  32 Glucinum.
  33 Yttrium.
  34 Zirconium.
  35 Thorinum.
  36 Cerium.

Difficultly fusible Bases.
  37 Iron.
  38 Manganese.
  39 Nickel.
  40 Cobalt.

Easily fusible Bases.
  41 Zinc.
  42 Cadmium.
  43 Lead.
  44 Tin.
  45 Bismuth.
  46 Copper.
  47 Mercury.

Noble Metals.
  48 Silver.
  49 Gold.
  50 Platinum.
  51 Palladium.
  52 Rhodium.
  53 Iridium.
  54 Osmium.

It is foreign to the object of this work, to enter into a minute description of these bodies; we shall therefore content ourselves with such a view of them, as may enable the general reader to form some idea of their properties; and to follow us, without much difficulty, in our subsequent remarks.

Of the Supporters of Combustion.—The five first bodies, *Oxygen, Chlorine, Bromine, Iodine,* and *Fluorine,* are usually termed supporters of combustion. They have some properties in com-

1. Oxygen, from οξυς, acid, and γενναω, to generate; from its property of forming acids. 2. Chlorine, from χλωρος, green; so called from its colour. 3. Bromine, from βρωμος, fetid; so called from its strong odour. 4. Iodine, from 'Ιοειδης, violet; from the colour it assumes in the gaseous state. 6. Hydrogen, from υδωρ, water, and γενναω, to generate. 8. Azote, from a privative and ζωη, life; from its being incapable of supporting life. 13. Selenium, from Σεληνη, the moon. 17. Chromium, from χρωμα, colour; so called from the beautiful colours of some of its salts. 18. Uranium, from ουρανος, the heavens. 19. Vanadium, from *vanadis,* a Scandinavian deity. 20. Molybdænum, from Μολυβδαινα, lead. 22. Titanium, from Τιτανος, calx. 23. Columbium, from Columbia, in America, where it was first found. 26. Lithium, from Λιθος, a stone. 29. Strontium, from Strontian, the name of a place in Scotland, where first found. 30. Baryum, from Βαρυς, heavy. 31. Aluminum, from Alumen, alum. 32. Glucinum, from Γλυκυς, sweet; from the taste o. some of its salts. 52. Rhodium, from 'Ροδον, a rose; from the colour of some of its compounds. 53. Iridium, from 'Ιρις, the rainbow; from the variety of colours assumed by some of its salts. 54. Osmium, from 'Οσμη, odour; from the strong smell emitted by some of its compounds.

mon ; though in other respects, and particularly, in their apparent relative importance in the economy of nature, they differ exceedingly. They are remarkable for the tendency they have, not only to combine with one another ; but with almost all the bodies below them in the table; and their union, particularly that of oxygen, is usually accompanied by the extrication of more or less, of heat and light ; and constitutes the well-known phenomenon, termed *combustion.*

(1) *Oxygen*, is one of the very few elementary substances, occurring naturally in the gaseous form ; in which form it is found in common air, in the proportion of about a fifth part. As the world at present exists, oxygen, perhaps, may be fairly considered as one of the most important, if not the most important, substance in it. From its proneness to enter into composition, it is con- stantly operating upon, and modifying, every thing. By far the greater proportion of mineral bodies, forming the crust of the earth, contain more or less of oxygen ; and in all plants, and ani- mals, it actually exists, as a constituent elemen- tary principle. In short, the properties of oxygen, stamp it as an element and subordinate agent, of the most important kind ; while the number- less contrivances which are observable in nature, to secure, or evade, or modify its operations, are truly extraordinary ; and exhibit some of the most unequivocal evidences of design, on the

part of their great Contriver, which we see among his works. Several of the most important of these contrivances, we shall have occasion to notice hereafter; but there is one of so curious and interesting a character, that it may be mentioned here, as an illustration of the above remarks. The nature and mechanism of the function of respiration will be explained elsewhere; it is sufficient for our present purpose to state, that, by means of a complicated apparatus, the blood is made to circulate through the lungs; where it is exposed to the influence of the oxygen of the atmosphere. For purposes beyond our comprehension, but probably, in part at least, with a view to the future creation of organized beings, the great Architect of the universe had willed that this principle should exist upon the surface of our globe in a gaseous state: when He created animals, He chose also to render them dependent upon oxygen for their existence; and He effects his object, not by bending this principle to his purpose, by altering its physical or other properties; not by obtaining it from water, or any of the innumerable compounds into which it enters, which according to our imperfect notions He might have more easily done; but, as if on purpose to display his power and design, He rigidly adheres to the properties, both mechanical and chemical, imparted to oxygen; and to these properties accommodates his future labours! The whole therefore, of the complicated and beautiful apparatus, connected with the respiration of animals, is most obviously designed and constructed, with reference to the properties of the oxygen of the atmosphere; and altogether, this apparatus affords one of the most striking instances of adaptation and design, presented to us in nature.

(2) *Chlorine,* in its elementary state is a gas, having all the mechanical properties of common air; but in this form it never occurs naturally. It exists however in great abundance in a state of combination, from which it may be readily obtained by easy chemical processes. One of the most abundant sources of chlorine is *common salt;* into which it enters in the proportion, of about 60 per cent. As compared with oxygen, chlorine is much less abundant and perhaps important; yet it is doubtful, if without chlorine, the present order of things could proceed. Take for example the familiar instance of common salt, above referred to. Let us consider the universal diffusion of this substance throughout nature—what the sea would be; or how animals could exist, without it.; let us consider these, and the numberless other operations, which this valuable compound more or less enters into, or influences; and we shall be able to form some notion of the part, chlorine bears in the economy

of nature. On the other hand, when we reflect, that were chlorine to be extricated from its state of combination, and made to exist, like oxygen, in a gaseous form, that it would instantly prove fatal to organized beings; can we fail to be struck with the very obvious design thus displayed, in rendering its quantity and combining powers such, as to keep it in a state of union; and by these means, to secure all its useful, without its deleterious properties?

(3) *Bromine*, and (4) *Iodine*, the next two substances, are found principally in sea water, and in marine productions. They appear to exist in very minute proportion, and always in a state of combination. *Bromine* under ordinary circumstances, is a deep coloured, red fluid, having a very strong and offensive odour. *Iodine* is a crystallized solid, volatile by a slight increase of temperature, and forming a beautiful violet vapour. Bromine and Iodine more nearly resemble chlorine, than oxygen, in their properties; though they differ materially from both; and their use in the economy of nature, is absolutely unknown. We may however observe, that *Iodine* has lately been much celebrated for its medicinal properties.*

* It may not be amiss also to notice, that the author of the present volume first employed the hydriodate of potash, as a remedy for goitre, in the year 1816; after having previously ascertained, by experiments upon himself, that it was not poison-

P.   I

(5) *Fluorine* has been rather inferred, than demonstrated to exist. It occurs principally in the mineral called *Fluor spar*, in a state of combination with lime ; and in this state it would seem to be harmless ; but in a state of purity, it is exceedingly deleterious. One of its most remarkable properties, is that of corroding glass.

Of Acidifiable Bases.—We pass on now to a very different class of substances, many of which, instead of having the power of supporting combustion in other substances, are themselves combustible. From their property of generally forming acids, when combined with the supporters of combustion, they have been denominated by Dr. Thomson, *acidifiable bases.* They are seventeen in number, and the first, and perhaps the most important, we have to notice is

(6) *Hydrogen.* This principle in its elementary state, exists as a gas, having all the mecha-

ous in small doses, as had been represented. Some time before the period stated, this substance had been found in certain marine productions ; and it struck the author, that burnt sponge, (a well-known remedy for goitre,) might owe its properties to the presence of Iodine, and this was his motive for making the trial. He lost sight of the case in which the remedy was employed, before any visible alteration was made in the state of disease ; but not before some of the most striking effects of the remedy were observed. The above employment of the compounds of Iodine in medicine was, at the time, made no secret; and so early as 1819, the remedy was adopted in St. Thomas's Hospital, by Dr. Elliotson, at the author's suggestion.

nical properties of common air. In this state it is exceedingly inflammable; and if mixed with oxygen, and if the mixture be exposed to heat, the two gases unite suddenly and violently with a loud explosion; while the result of the combustion is *water*. With the other supporters of combustion, hydrogen forms compounds, more or less acid. Hydrogen is the lightest body known, and under the same bulk therefore, contains less matter than any other body. It does not exist naturally in a separate state, but always in combination; and by far most generally and abundantly in combination with oxygen, in the form of water. Hydrogen ranks perhaps next to oxygen in importance; at least as far as organized beings are concerned; since, like oxygen, it constitutes one of the elementary principles of which they are formed. It differs however, remarkably from oxygen, in not being in its elementary state necessary to the existence of organized beings; indeed hydrogen is actually incompatible with the existence of animals, if not of vegetables; and its properties as an element, have evidently been sacrificed to its properties as a compound; that is to say, to its properties as *water*. Hence we have to admire the happy adjustment of the quantities of the two elements to each other, so that the oxygen shall predominate; an adjustment that can scarcely be explained on any other supposition

than that of design; for any other cause, as
chance, would have been quite as likely to have
produced an excess of hydrogen as of oxygen;
or at least any thing but the exact proportions
required. Lastly, it may be remarked, that to
the relative proportions of oxygen and hydrogen
existing on our globe, more than perhaps to any
other subordinate cause, the present order of
things owes its stability. For the proportions of
these principles are so happily adjusted; and
all the numerous operations dependent upon
them are, in consequence, so firmly established;
that no material change can possibly happen to
any part, from an internal cause ; but if changed
at all, the whole must be changed from without.

(7) *Carbon*, or charcoal, is a substance too
well known in its ordinary state to require description. In its crystallized and pure state,
carbon is found to constitute the *diamond*, the
hardest and most brilliant body in nature—a
circumstance which certainly could not have
been anticipated; but which affords a most
striking instance, of the effects produced, by the
different modes in which molecules of the same
matter may be aggregated. Carbon, perhaps
more than any other principle, may be considered as constituting the staminal, or fundamental element, entering into the composition of
organized beings. This is particularly the case
in principles from the vegetable kingdom ; which

owe their peculiar character, essentially to carbon ; and their endless varieties, to differences
in its quantity ; or to the modifying influence of
the hydrogen and oxygen, with which it is associated. In animal substances, carbon exerts a
similar influence ; but its effects are materially
modified by the presence of another staminal
principle, to be presently considered. Carbon,
in some state or other, exists in considerable
quantities upon the surface of our globe ; but
apparently by no means in so large a proportion,
as oxygen and hydrogen. Exclusively of that
actually involved in the composition of organized
beings, carbon is met with nearly pure in large
quantities, in particular districts, in the well-
known form of *fossil coals;* but it occurs in far
greater proportion in combination with oxygen,
in the form of *carbonic acid;* which carbonic
acid in union with lime, constitutes common
chalk and *lime stone;* two of the most abundant
minerals in nature.* Carbon in its elementary
state, is a very inert substance ; and is scarcely
liable to be affected by, or to affect organized
beings ; but with hydrogen and oxygen, it forms

* In order to give some idea of the proportion in which carbon exists in different common substances, it may be observed,
that a pound of charcoal is equal to, and is contained in rather
more than, two pounds of sugar or flour, and eight of potatoes
or limestone ; so that a mountain of limestone, contains the
essential element, of at least, an equal bulk of potatoes ; and of
a forest, that would amply cover many such mountains.

gaseous compounds of great activity, which prove instantly fatal to animals respiring them. Such effects, however, appear to be obviated by a beautiful expedient, to be noticed hereafter. In the mean time it may be observed, that though the compound of carbon and oxygen, (*carbonic acid*,) is by innumerable processes constantly forming around us in enormous quantities; by some compensating means, it disappears as fast as it is formed; so that the atmosphere, which without this provision, would probably before now, have become contaminated by carbonic acid to an extent fatal to animal life, barely contains traces of it.

(8) *Azote*, or *nitrogen*, is one of the very few elementary principles which exist naturally in an uncombined state. It constitutes about 4-5ths, or 80 per cent. of common air; the rest being principally oxygen. The great bulk of this principle in existence is confined to the atmosphere; or to animal substances, of which it forms a constituent element: and it enters very little into natural mineral productions. In its pure state, azote is remarkable for its negative properties; that is to say, for the difficulty with which it enters into combination with other matters. Thus, it is neither combustible, nor a supporter of combustion; is neither acid, nor alkaline; possesses neither taste, nor smell; nor does it directly combine with any known sub-

stance. Yet when made by peculiar management to unite with oxygen, hydrogen, or carbon,
azote forms some of the most energetic compounds we possess : thus, *mixed* with oxygen, it
forms atmospheric air, as before observed ; *united*
with oxygen, it forms *aquafortis*, the most corrosive of liquids; *united* with hydrogen, it forms
the *volatile alkali*, or ammonia, likewise an energetic compound, but of an opposite nature ;
while *united* with carbon and hydrogen, it forms
prussic acid, the most virulent poison in existence ! Azote may be considered as constituting
the characteristic element of animal substances,
and as imparting to them their peculiar properties ; in this point of view therefore, it is a principle of very great importance. Moreover, the
above mentioned negative properties of azote are
evidently of a primordial kind, and seem to have
been formed with reference to future creations ;
which have all been most carefully and rigidly
adapted to them. Thus, had the properties of
azote not been negative, those of its most important compound, atmospheric air, could not
have been negative ; and atmospheric air might
have been acid or alkaline ; or have possessed
odour or colour ; either of which circumstances
would have been incompatible with the present
order of things.

(9) *Boron*, and (10) *Silicon*, the next two
substances obtain their names from *borax*, and

silex, the natural productions in which they exist, in a state of combination. Borax is a saline production, chiefly found in certain lakes in Thibet and China. Boron, the elementary substance obtainable from it, is a deep brown powder, possessing neither taste nor smell; but highly inflammable, at a temperature below a red heat. The resulting compound thus formed with oxygen, is boracic acid. Silicon, is the elementary basis of silex or common flint; one of the most abundant minerals in nature. Silicon is a brown powder very similar to boron in appearance; and like it, inflammable under certain circumstances. By combustion, silicon combines with oxygen, and is converted into silex; which many chemists consider as an acid. Boron and silicon do not exist naturally; but have been formed by elaborate chemical processes, in small quantities only. They seem to be more nearly allied to carbon, in their properties, than to any other elementary product. Borax exists in very small quantities; and its use in the economy of nature, is not apparent. Silex, on the other hand, is a most important production; and in its hardness, insolubility, and other refractory properties, we recognise a substance admirably adapted for the purpose, to which it has evidently been designed; viz. that of constituting the stamina, or ground work, as it were, of our globe; and which could not be

withdrawn without subverting the whole. Silex
is found in small quantities, both in plants and
in animals; but does not, like hydrogen, oxygen,
carbon, and azote, form a constituent element of
organized beings.

(11) *Phosphorus*, under ordinary circumstances, is a pale amber-coloured substance, very
like wax in appearance; but so exceedingly
combustible, that it cannot be heated, much less
melted, in the open air, without immediately
taking fire: the product of the .combustion is
phosphoric acid. Under these circumstances,
as may be supposed, phosphorus does not exist
naturally, but is obtained by an elaborate process from various products into which it enters;
as for example, from *bone earth,* or the earthy
basis of the bones of animals; and from other
saline compounds. It exists also in the mineral
kingdom in certain districts, in considerable
quantities; though upon the whole, it is not an
abundant principle. Phosphorus affords another
beautiful instance, in which the design has been
directed to the properties of the compound,
rather than to the element itself. The *phosphate
of lime,* or bone earth, was apparently the thing
wanted, to constitute the bony skeleton of animals; and accordingly, to the properties of this
compound, the properties of the element seem to
have been sacrificed. Neither lime itself in
mass, nor any of its mineral compounds, appear

to be adapted for forming a constituent principle of a living organized being. It was necessary, therefore, to have a connecting medium, or link, that should unite organization with the mineral constituent ; and phosphorus admirably accomplishes this object. Accordingly, we see, that organization goes on in conjunction with lime in the bones of animals, through the medium of this element, quite as readily as in other parts of their system : whereas, when phosphorus is absent, as in shells, and in other deposites of carbonate of lime ; the carbonate of lime is extravascular, and seems to form no part of the living system. There are also other important offices, which this principle evidently performs in the animal economy ; some of which we shall have occasion to refer to, hereafter.

(12) *Sulfur.* This well known substance is one of the very few that exist naturally in an elementary state. It is a very abundant, and probably, important, principle in the economy of nature ; as it not only exists in large quantities in the mineral kingdom, but in a greater or less proportion, in almost all animal, and in many vegetable products. Its uses, however, at present, are very imperfectly understood. Sulfur combines with hydrogen, and forms a very deleterious gaseous compound. Its combinations with oxygen are generally acid, and very active in their concentrated form ; but not poisonous.

(13) *Selenium*, the next substance, is found in very minute quantities, generally associated with sulfur; which in its properties it somewhat resembles: or rather, perhaps, it appears to constitute the connecting link between sulfur, and the metals. The uses of selenium in the economy of nature are unknown; but we shall have occasion hereafter, to refer to its compound with hydrogen; which is even more deleterious, than the compound of sulfur with this element.

(14) *Arsenic*, in its pure state is a metalloid, or imperfectly metallic substance, having much the appearance of polished steel. In the form in which it is popularly known, as *white* arsenic, it is combined with oxygen; and constitutes one of the most virulent of poisons. Arsenic exists in certain minerals in considerable quantities; but seems, in every form, to be incompatible with organic life.

(15) *Antimony*, is usually found, associated with sulfur; the compound it forms with which, was for a long time, considered as the metal itself. In its pure state, antimony has a bluish-grey colour, and possesses considerable metallic splendour; but in this form, it seldom occurs in nature. The compounds of antimony are active medicinal agents; and some of them are much employed for that purpose.

(16) *Tellurium*, (17) *Chromium*, (18) *Uranium*, (19) *Vanadium*, (20) *Molybdænum*, (21) *Tungsten*,

(22) *Titanium*, and (23) *Columbium*, the next eight substances, are metals, for the most part obtained by elaborate processes, from rare mineral productions. The most important, as well perhaps as the most abundant, of these substances, is chromium ; the compounds of which, from the splendour of their colours, have been lately much employed in the arts. Like selenium, arsenic, and antimony, these metals all combine with oxygen, &c. and form compounds, possessing many of the characters of acids. It may be remarked of all these substances, that at present their use in the economy of nature is quite unknown to us.

Of Alkalifiable Bases.—The next thirty-one bodies have been denominated by Dr. Thomson *alkalifiable bases;* from their property of forming compounds partaking more or less, of the character of those of the first subdivision, or family, termed *alkalies.* Dr. Thomson has subdivided the alkalifiable bases into five families, the designations given to which, sufficiently mark their character—viz. the *alkaline* bases ; the *earthy* bases ; the *difficultly fusible* bases ; the *easily fusible* bases ; and the *noble metals.*

Of the Alkaline Bases.—(24) *Potassium,* and (25) *Sodium,* are the metallic bases of the two well-known alkaline substances, *potash* and *soda;* which are compounds of these metals with oxygen. Such, however, are the powerful affinities of the metallic bases for oxygen, that they no where exist naturally upon the surface of our globe. The same may be also remarked of potash and soda; the powerful alkaline properties of which, prevent them from existing separately. In this respect, the compounds these metals form with oxygen, present a striking contrast with the compounds they form with the analogous principle, chlorine; the compounds of potassium and sodium with chlorine, (the latter of which constitutes common salt,) are remarkable for their permanent character; and for the little tendency in general, which they have, to enter into a further state of combination. Besides their remarkable avidity for oxygen, potassium and sodium possess some other unusual properties. Potassium, for example, is so light, that were it compatible with water, it would swim on the surface of that fluid; a circumstance we can hardly imagine to happen with a metal. Potash and soda, in all their forms, are most important principles; and evidently are necessary to the existence of the present order of things, both mineral and organized: for there are few organized beings, that do not contain more or less of them; especially of soda. Potash is found more particularly in plants; but exists also in animals: while the universal presence of soda in animals, in the form of common salt, has been already referred to, and is generally known. These alkalies present us with a beautiful instance of adaptation, for the purposes which they seem destined to fulfil, in the operations of nature. Had they been solids; or had they formed solid compounds, like many of the preceding principles; they would have been totally unfitted for their peculiar office; that is to say, for forming a constituent element of the fluids of organized beings.

(26) *Lithium*, is the metallic basis of the alkaline substance termed *lithia*. This newly discovered substance is intermediate in its properties, between the alkalies and the earths, to be next considered. It has hitherto been met with, in some rare minerals, in small quantities only.

(27) *Calcium*, the metallic basis of *lime*, can be obtained only by a troublesome and difficult process; and, of course, does not exist naturally. It is a white metal like silver, and by union with oxygen, is readily convertible into lime. This well known principle exists in the greatest abundance in nature; not as quick-lime; but united with carbon and oxygen, in the form of common lime-stone, marble, &c. The great importance of lime in the economy of nature, is too obvious to require notice; and it is only necessary to revert to the fact, that this earth is one of the very few mineral productions, capable

of forming a part of a living organized being ; at least in any quantity. This earth, as formerly noticed, constitutes with phosphorus and oxygen, the basis of the bones of animals; and with carbon and oxygen, all the endless variety of shells, and similar products. Thus the properties of lime, furnish another striking instance of adaptation to a particular purpose. The compounds of potash and soda are all very soluble in water, and hence are chiefly confined to the *fluids* of animals; in which their presence is indispensable. But a solid frame work, or skeleton, was necessary to the existence of the more perfect animals ; and as this could not be formed from the soluble potash or soda; the introduction of another mineral substance, possessed of the requisite properties, was necessary. Now lime, some of the compounds of which are *solid*, and some *fluid*, is admirably adapted for the purpose ; and lime accordingly has been chosen : the lime is carried, in a state of solution, to the spot where it is required, and is there converted into a *solid ;* while by the same agency, when necessary, this solid is again converted into a fluid and removed !

(28) *Magnesium*, is the metallic basis of the well-known earth, called *magnesia.* It is said to resemble calcium in its properties ; and like that principle, does not exist naturally, at least upon the surface of our globe. Magnesia, though oc-

curring most abundantly in nature, and entering very largely into the composition of rocks; never, like lime, constitutes masses of great extent, in the same simple state of combination; that is to say, there are no mountains of magnesia; as there are of chalk and of limestone. Magnesia, even more decidedly than the three preceding mineral substances, seems to be necessary to the existence of organized beings; as there does not appear to be one, in which traces of this earth are not met with, generally associated with phosphorus. Its uses, however, are less obvious than those of the three other substances, and indeed may be said to be unknown: though there is reason to believe, that it is most intimately connected with the vital operations of organized beings.

(29) *Strontium*, and (30) *Baryum*, the metallic bases of the two alkaline earths, *strontia*, and *baryta*, are allied to calcium and magnesium in some of their properties; but differ exceedingly from them in others. Their combinations with oxygen exhibit still more decidedly alkaline powers, than those of either calcium, or magnesium; and in consequence, like them, they only exist in various states of combination; and most usually, with carbon and oxygen; or with sulfur and oxygen. Compared with lime and magnesia, strontia and baryta exist but sparingly; and neither of them has any thing to do with

organization; indeed, many of the combinations of baryum are virulent poisons.

Of the earthy Bases. (31) *Aluminum,* is the metallic basis of the earth *alumina;* the characteristic ingredient of the well known salt, called *alum.* The metallic basis, like the preceding, no where exists; but alumina, the compound of aluminum and oxygen, is one of the most abundant productions of nature; and constitutes an ingredient, in by far the greater number of rocks and soils, upon the surface of the globe. The different kinds of clay, also, from which bricks, earthenware, &c., are formed, consist chiefly of this earth, in different states of purity; so that it is a substance of great utility and importance. Alumina appears to have nothing to do with organization; at least, it is not known to form a necessary constituent of any organized being, either vegetable or animal; though it is in constant communication with organized beings; and appears to be almost necessary, in some indirect way, to their existence. This fact is very remarkable; for as the earth does not appear to be poisonous, it could scarcely have been so completely excluded from living bodies, except by some design beyond our comprehension.

(32) *Glucinum,* (33) *Yttrium,* (34) *Zirconium,* and (35) *Thorinum,* the four next elementary principles are the metallic bases of substances, usually considered as possessing the characters of

P.   K

earthy bodies, and denominated *Glucina, Yttria, Zirconia,* and *Thorina.* They all appear to exist very sparingly in nature ; and are only met with in some rare minerals. Glucina has been hitherto met with only in the precious stones, denominated the *emerald,* the *beryl,* and the *euclase;* yttria and thorina in some rare Swedish and Norwegian minerals ; and zirconia in the *jargon,* or *zircon* from Ceylon, and in the *hyacinth.* These earths more nearly resemble alumina, than any other substance.

(36) *Cerium* is a metal very little known, and has hitherto been obtained, in minute quantities only, from some rare minerals occurring in Sweden and in Greenland.

Of the difficultly fusible Bases. (37) *Iron,* one of the most important, is also one of the most abundant principles in nature. It is met with occasionally in the metallic state ; but most generally, it is found mineralized in various ways; and can only be obtained pure, by an elaborate process. Iron exists in minute quantities in almost all vegetable and animal products, particularly in the blood ; though its mode of combination, as well as its precise use, are quite unknown. Iron may justly be considered as the most useful of all the metals ; and the one, that has perhaps contributed more towards the civilization of mankind, than any other. To form some idea of its use, we have only to reflect, what would happen if it were annihilated. What substitute could be found for it, in all the numerous instances in which it contributes to the wants, or to the comforts, of mankind; particularly through the medium of tools, of almost every one of which, it constitutes the essential material. In short, when we contemplate all the circumstances connected with this metal ; its abundance, the manner in which it is mineralized, and the occasion which it thus gives to human ingenuity to extract it from its ores ; its wholesomeness, (for many of the metals are poisonous) ; its properties, particularly its extraordinary tenacity, its strength, its property of welding, of being converted into steel, and in this form of being tempered to any degree of hardness we choose ; its magnetic properties, &c.,—when we contemplate all these circumstances, it is impossible not to be struck with such varied usefulness ; and to consider iron, not only as an article evidently designed for the benefit of man ; but as the instrument, by which he should conquer, and govern, the world ; and thus be enabled to place himself, where it was evidently intended he should be, at the head of the creation.

(38) *Manganese*, somewhat resembles iron in a few of its properties. It may be obtained from its ores by an elaborate process ; but in this form it is little known or used. Manganese exists in minute quantities in certain mineral waters; and

in a few animal products. The combinations of this metal with oxygen are employed in the arts; the chemist also frequently procures oxygen for his experiments, from the ores of manganese. Though much diffused, manganese is not a very abundant metal, at least compared with iron; and its uses in the economy of nature are apparently much less important.

(39) *Nickel*, and (40) *Cobalt*, are two metals somewhat resembling each other in a few of their properties; and their ores are often associated in nature. It is remarkable also, that they are both generally found combined with iron, in those bodies, which occasionally fall from the atmosphere; and which are considered as of meteoric origin. Like iron also, both these metals are capable of becoming magnetic. Cobalt is used in the arts, and is the basis of the blue colour upon our earthenware; but neither this metal, nor nickel, are to be compared with iron in point of utility: nor are they very abundant productions.

Of the easily fusible Bases. (41) *Zinc.* (42)*Cadmium.* These two metals are generally associated in nature, and somewhat resemble each other in their properties; but cadmium is comparatively much less abundant than zinc, and has been only recently discovered. Zinc is a metal easily fusible; of a bluish white colour; and of a lamellated brittle texture: though by peculiar management, it may be rendered malleable. It is an ingredient in the well-known metal, *brass;* and in this form is much used, and is of considerable importance.

(43) *Lead.* This well-known metal is not found in its metallic state, but its ores are very abundant ; and most of the lead of commerce, is extracted from the mineral, called *galena,* which is a compound of lead and sulfur. The general properties of lead, and of its compounds, render it of considerable importance ; but its poisonous properties are a considerable draw-back to its usefulness. Why lead, and other mineral matters, should have been constituted poisonous, is a question beyond our reach ; and all we can at present venture to state on this and on similar points is, that it is not actually necessary, that man should make use of lead or other poisons ; and that he may, if he chooses, avoid their deleterious properties.

(44) *Tin.* This useful metal has been employed by man from the most remote antiquity ; though it no where exists naturally in its metallic state ; but usually in conjunction with oxygen. Tin is not a very abundant metal, being apparently confined to a few localities only ; one of the most noted of which, is Cornwall. It is much used in the arts ; and hence is of considerable importance.

(45) *Bismuth,* occurs in nature, both in the

metallic state, and in various states of combination. It has a reddish white colour; a lamellated brittle texture; and is easily fusible. Bismuth is not a very abundant metal; nor is it much employed.

(46) *Copper*, occurs in nature in the metallic state; but much more frequently mineralized, especially with sulfur. The valuable properties of copper, both in its pure and mixed state, render it of considerable importance; and it is, in consequence, much employed in the arts. With zinc, it constitutes *brass;* with tin, *bell-metal;* both well-known compounds. Copper has been lately said to exist in very minute quantities in organic nature; but whether as an accidental, or as an essential ingredient, is not known. The compounds of copper are poisonous; but these poisonous properties, like the poisonous properties of lead, can be easily obviated, and guarded against.

(47) *Mercury.* This well-known fluid metal occurs in the metallic state; but more frequently mineralized, especially with sulfur. Its importance in the arts, and as a medicinal agent, need scarcely be mentioned here. The fluidity of mercury, presents a beautiful instance of the endless diversity of nature; and adds much to its importance, and usefulness. Mercury exists in considerable abundance; though much less

so, than many of the preceding elementary principles.

Of the noble Metals.—(48) *Silver,* and (49)
Gold, and their uses, are too familiar to require
enumeration.   They are both met with in the
metallic state; but silver also occurs mineralized.
So unimportant a part do they seem to perform
in the economy of nature; that if they were
annihilated, it is probable that the world would
go on, just as well without them.   How different
in these respects from iron; and how much less
therefore intrinsically valuable!   Independently
of their beauty; the only really valuable properties of silver and gold, are the difficulty with
which they are acted on by heat, and other extraneous agents; properties, which if they were
more abundant, would render them well adapted,
for a great many useful purposes.

   (50) *Platinum,* (51) *Palladium,* (52) *Rhodium,*
(53) *Iridium,* and (54) *Osmium,* are metallic
substances usually found associated in small
quantities, chiefly in certain districts of South
America; but recently also, in the old world.
Platinum, the most abundant and important of
them, is the heaviest body in nature.   It is acted
on with difficulty by most ordinary agents; but
it may be welded by heat — properties which
render it exceedingly valuable for many purposes; and make us regret, that it is not more

abundant.   Palladium somewhat resembles platinum in its characters ; but occurs in less quantity.   The other three metals exist in very minute quantities ; and their properties are very imperfectly known.

We have thus taken a summary view, of the different elementary principles met with, upon the surface of our globe ; and of their leading properties.   The subject in the next place, to be considered, are the compounds, which these principles form with one another.

Section II.

General Remarks upon Chemical Compounds.

The number of chemical compounds is so great that an attempt to enumerate them here, would be quite out of place ; we shall therefore content ourselves with stating, as briefly as possible, the general principles upon which these compounds are formed.

We have already described many of the more remarkable compounds, when speaking of simple bodies ; and in subsequent parts of this volume, we shall have occasion to allude to others. In speaking of simple bodies, we showed that by far the greater number of them occur in the metallic state; and are incapable of existence upon the surface of our globe, on account of the tendency they possess to enter into combination, particularly with oxygen. It would seem also, from the intensity of the properties, and the general incompatibility of the simple bodies with the present order of things; that their compounds, rather than themselves, were the objects the Author of nature had in view. Hence perhaps we are more immediately interested in the character of the compounds, than in that of the elements themselves. Of the general nature of these compounds, the following observations, taken chiefly from Dr. Thomson's work on chemistry, will serve to convey some idea to the general reader.

The COMPOUNDS which bodies form with one another, are either PRIMARY, or SECONDARY. By PRIMARY COMPOUNDS, are usually understood those which are formed by the combination of two or more simple bodies with each other; while by SECONDARY COMPOUNDS, are meant the compounds formed by the union of the primary compounds with each other.

The PRIMARY COMPOUNDS naturally divide themselves into three grand classes; viz. *acids; alkalies*, or *bases;* and *neutrals;* on each of which we shall make a few remarks.

Of Acids. Formerly it was considered as requisite, that bodies, in order to belong to the class of acids, should have a *sour* taste, should be soluble in water, and should have the property of reddening vegetable blue colours; and these properties do indeed belong to some of the most common, and powerful acids. But there are various acids which have no taste; which are not soluble in water; and some, which are incapable of altering the colour of the most delicate vegetable blues; hence the term *acid*, as at present employed by chemists, is understood to denote a substance which has the property of combining with, and neutralizing, alkalies or bases. The celebrated Lavoisier endeavoured to prove, that oxygen constitutes an essential ingredient of all the acids; but later observations have shown, as already stated, that not only oxygen, but the analogous principles, chlorine, bromine, iodine, and fluorine, are also capable of forming acids, by uniting with several of the acidifiable bases. Still more recently, certain compounds of cyanogen, (a primary compound of carbon and azote), of sulfur, of selenium, and of tellurium, with the acidifiable bases, have been ranked among the acids; so that the acids at present known, may be divided into nine classes, viz. *oxygen acids, chlorine acids, bromine acids, iodine acids, fluorine acids, cyanogen acids, sulfur acids, selenium acids,* and *tellurium acids.*

The *oxygen acids* are more numerous, and better understood, in general, than the other classes; they may be subdivided into two kinds; those with a single base; and those with a compound base. The acids with a single base, amount to between thirty and forty; and include most of the best known and most important of those used in chemical processes, and in the arts; such as *carbonic acid, sulfuric acid, phosphoric acid, nitric acid,* &c. The oxygen acids with a compound base are chiefly derived from the vegetable or animal kingdoms; they are still more numerous than those with a single base, the number at present known, amounting to upwards of sixty; as instances may be mentioned the *tartaric acid,* the *citric acid,* the *malic acid,* the *lithic acid,* &c.

The *chlorine acids* are perhaps as numerous as those with a single base, containing oxygen; but they have been much less studied, and are, consequently, much less understood. One of the most familiarly known belonging to this class, is the *muriatic,* or *hydrochloric acid;* which is composed of chlorine, united with hydrogen: and here may be noticed a remarkable circumstance, before alluded to, that not only chlorine, but all the other allied principles, when they combine with hydrogen, form powerful *acids;* while the compound of oxygen with hydrogen, is *water;* a substance altogether dissimilar. Such is the wonderful and inexplicable nature of chemical combinations!

The *acids* containing *bromine, iodine,* and *fluorine,* are still less satisfactorily known, than those containing chlorine. As just observed, the acids formed by these different principles with hydrogen, viz. the *hydrobromic,* the *hydriodic,* and the *hydrofluoric acids,* possess the most decided properties, and are best understood.

The *cyanogen acids* are numerous and important, as most of them are poisonous; thus the compound of cyanogen and hydrogen, (analogous to those above mentioned), is the *hydrocyanic,* or *prussic acid;* one of the most virulent poisons in nature, and instantly fatal to organic life in every form.

Of the remaining acids, the *sulfur acids,* the *selenium acids,* and the *tellurium acids,* we know very little. Those with which we are at present best acquainted, are analogous to the preceding acids, and are formed by the union of the different principles with hydrogen. These acids were formerly known under the names of *sulfuretted, seleniated,* and *telluretted hydrogen;* but some chemists have now given them new names, conformably to the above nomenclature.

Of Alkalies and Bases. Bodies of this class, are, as we have seen, like the acids, composed of different elements, and particularly of certain metals, combined with oxygen, chlorine, &c.; but usually in less proportions than in the acids. Hence the alkaline bases are as numerous as the acids, and may be divided, in a similar manner, into *oxygen alkalies, chlorine alkalies*, &c. Of these, the *oxygen alkalies* are by far the best known, and most important; and they may, like the oxygen acids, be subdivided into two kinds: viz. those with a single base, and those with a compound base. The alkalies with a single base, include all the well known common alkaline bodies, *potash, soda, lime, baryta*, &c.; while the alkalies with a compound base, are chiefly from the vegetable kingdom; and comprehend the newly discovered alkaline matters, so successfully introduced into medicine; such as *quinine*, from bark, *morphine*, from opium, &c., the composition of which at present is not well understood. *Ammonia*, or the *volatile alkali*, may perhaps be referred to this class of alkalies; though its composition as consisting of hydrogen and azote only, without oxygen, may be considered as constituting an exception or anomaly.

The other *alkaline* bodies into which *chlorine, iodine*, &c., enter, are very little known; and some, perhaps, may be even inclined to doubt their existence.

Of neutral Compounds. These are arranged by Dr. Thomson under seven heads, the mere naming of which, will probably be all that is required, to convey to the general reader, a sufficient notion of their nature. They are *water, spirits* or *alcohol, ether, ethal*, (a peculiar oily substance obtained from spermaceti) *volatile oils*, *fixed oils*, and *bitumens*.

Such is a summary of the PRIMARY COMPOUNDS, and of the principles upon which they have been most recently arranged. We come now to consider briefly

The SECONDARY COMPOUNDS ; or those formed by the union of the primary compounds. As the neutral primary compounds, (if we except water), enter into few combinations, it is obvious that the SECONDARY COMPOUNDS must consist chiefly of substances formed by the union of the other two general classes of bodies ; namely, of *acids* and *alkalies*. These SECONDARY COMPOUNDS are usually denominated SALTS ; they constitute a very numerous and most important class of bodies ; and, as resulting from the mutual union, and saturation, of all the different principles capable of combining with each other, they of course are more abundant than any other bodies ; indeed, the surface of our globe may, in a great measure, be considered as made up of them. The term *salt*, was originally confined to common salt ; but by a singular fate, this body, as being composed of chlorine and sodium only, is now excluded from the class of salts : salts being, as we have just said, considered by chemists, to be formed by the union of acids and alkalies only. As there are nine classes of acids, of course there must be as many classes of

salts; of these, the *oxygen acid salts* are by far
the best known, and the most important; and,
indeed, this class includes the greater number of
those salts employed by chemists, or in the arts.
If the salts be arranged according to their
bases, which perhaps upon the whole, in the
present state of our knowledge, is the best mode
of arranging them, they will be found to constitute upwards of fifty genera; and if we consider that each of these genera includes, in most
cases, a great number of species; we may form
some idea of the wonderful variety of bodies
existing in nature; and with the properties of
which, the chemist is required to be conversant.
Familiar instances of the oxygen acid salts are,
nitre, common chalk, gypsum; various metallic
salts, as the *white, green,* and *blue vitriols,* &c. &c.

Of the *chlorine,* and the *other classes of salts,*
very little is known, and this little, is chiefly
confined to the salts, composed of these principles and of hydrogen. The hydrochloric or
muriatic acid combines with ammonia, and
forms the well known compound *sal-ammoniac,*
a salt supposed to be a true *hydrochlorate,* or
muriate. But this is the only instance known;
and in all other analogous instances, the hydrogen of the hydrochloric acid, and the oxygen
of the base, unite to form water, which is separated, or separable; and thus the chlorine and
the metallic base are left in union by themselves,

in the state of a chloride. This is the case, for instance, with *common salt;* which, as we before said, is in reality a *chloride* of *sodium;* that is to say, a simple compound of chlorine, and the metal sodium. Similar remarks appear to be applicable to the other analogous compounds. It must be confessed, however, that our knowledge with respect to all these matters is at present in a very unsatisfactory state ; and is probably destined, at no very distant time, to undergo a complete revolution.

SECTION III.

Of the Laws of Chemical Combination.

As the following remarks upon the laws of chemical combination, can scarcely be so given, as to prove a source of interest to the general reader ; he is desired to pass them over, and to turn to the last section of the present chapter; where, he will find a recapitulation of the leading facts ; together with the evidence which they furnish in favour of design ; and of the wisdom, and power, of the Creator.

In the preceding chapter, we briefly considered the arguments, which induced us to adopt the hypothesis, that *all gaseous bodies under the same pressure, and temperature, contain an equal number of self-repulsive molecules:* we have now to point out, some of the important consequences, to which this hypothesis naturally leads.

It seems to be satisfactorily established, that bodies, in their gaseous state, combine both *chemically*, and *cohesively*, with reference to their volumes : that is to say, that the same volume of a gas, always combines with either precisely a similar volume of the same, or of another gas ; or with some multiple, or submultiple, of that gas ; (in other words with twice, or thrice, or half, or a quarter, as much, &c.) but not with any intermediate proportion ; and further, that the resulting compound, always has reference by volume, to the original volumes of its constituent elements. Let us take water for example. Water has been shown to consist of one volume of oxygen gas, and two volumes of hydrogen gas ; and so invariably, that we cannot suppose water to be formed of any other proportions of these elements. It has been also shown, that the resulting water, if in the state of steam, occupies exactly the space of two volumes ; so that one volume has disappeared. Now let us consider attentively, what must have happened during these changes. One volume of oxygen gas has contributed to form two volumes

P.   L

of water; which two volumes of water, according to our hypothesis, must consist of twice the number of self-repulsive molecules contained in the one volume of oxygen; yet every one of these molecules must contain oxygen; because oxygen is an essential element of water: it follows, therefore, irresistibly, that every self-repulsive molecule of oxygen, *has been divided into two;* and consequently, must have originally consisted, of at least two elementary molecules; somehow or other associated, so as to have formed only one self-repulsive molecule. This conclusion, which seems to flow inevitably from our premises, is most important, as we shall see immediately; and enables us to throw no small light upon many points ·deemed obscure. In the mean time, let us consider briefly the nature of the compound self-repulsive molecule of oxygen.

We endeavoured to show in the previous chapter, that every ultimate molecule of matter must possess two kinds of polarity; which, for want of better terms, we denominated the *chemical,* and the *cohesive;* and that these polarities bear the same relations to each other, as electricity and magnetism; in other words, that, like these forces, the chemical and cohesive polarities, exist at right angles to each other. Hence if A, and B, be supposed to be two molecules of oxygen, of which E*e*, E*e* represent the chemical axes, and MM, *mm*, the equatorial or cohesive diameters; it is evident, that these two molecules may be supposed to combine in two ways, either E, to *e*, chemically; or M, to *m*, cohesively; but the cohesive combination, of course, is most probable, *from the similar nature of the molecules.** Every self-repulsive molecule of oxygen, therefore, as it exists in a state of gas, must consist of *at least* two molecules, united to each other *cohesively*, and acting as a single one. Whether the self-repulsive molecule of hydrogen be double or not, cannot be inferred from the composition of water, as above stated; but this may be demonstrated, from other compounds, into which hydrogen enters. Thus muriatic acid gas is composed of one volume of chlorine,

* The general and strong analogy, if not identity, in all respects except *direction*, between the axial and the equatorial forces, has been already alluded to, and is exemplified, by the striking resemblance, between electricity and magnetism. We have seen also, that in the crystallized state, similar molecules probably combine *chemically*. Hence, although the rule stated in the text be true, that *similar* molecules only combine *cohesively*; yet there may be, and probably are, instances in which they combine *chemically*. For the same reasons, *dissimilar* molecules may also occasionally combine *cohesively*. It is probable that such states of combination might be readily detected by the optical properties, or by some other peculiarity in the physical properties of bodies, if in a crystalline form; but by no other known means. Do not some of the phenomena of *Dimorphism*, that is to say, the property which the same body occasionally possesses of assuming different forms, depend upon these changes?

and one volume of hydrogen ; which unite without any condensation, and form two volumes of muriatic acid gas : now, in this case, it is evident that not only the self-repulsive molecule of hydrogen, but also that of the chlorine, must be double at least, like the molecule of oxygen above mentioned ; and the same might be shown with respect to the other gaseous bodies.

We have said above, that the self-repulsive molecules of oxygen and of hydrogen, are *at least* double ; but the probability is, that they are in reality much more compounded ; as the following observations will show. The self-repulsive molecule of water, on entering into combination, is often found to be divided into two, or three, (perhaps more,) parts. Now as we cannot admit the division of an ultimate molecule, or *atom ;* we must of course conclude, that the molecules of oxygen and of hydrogen, are much more compounded, than as above represented ; and must each of them contain at least, three component, or sub-molecules. Hence the self-repulsive molecules of water will consist of at least nine component sub-molecules, (viz. three of oxygen, and six of hydrogen), which we may suppose to be associated—in the first place, the hydrogen with the oxygen, *chemically;* and afterwards, the three sub-molecules of water with one another *cohesively,* so as to constitute one spheroidal molecule ; in a manner, that with

a little ingenuity, it would, perhaps, not be difficult to represent mechanically.*

Precisely the same laws of union, may be supposed to prevail among the molecules of bodies
themselves, as they actually exist around us.
Thus, let us take the crystal of oxalic acid, as
an instance for illustration. This acid is composed, according to the present language of
chemists, of two molecules of carbon, and three
of oxygen, which by combining, form the acid;
while, to complete the compound molecule, and
to adapt it for crystallization, three molecules of
water are required to be somehow associated,
with each of the molecules of the acid. Now in
this case, we suppose, that the two molecules of
carbon, (each of which is perhaps already made
up of several sub-molecules,) are associated
together into one symmetrical super-molecule;
that the three molecules of oxygen, associated in
a similar manner, are then combined *chemically*
with the super-molecule of carbon, and thus
form by their union a molecule of oxalic acid;
finally, that the three molecules of water are
associated into one super-molecule, which unites

---

* When bodies, as, for example, water, are subjected to
intense degrees of heat, it is not improbable that in many instances the self-repulsive molecules are, more or less, separated
into their constituent sub-molecules; in which case, of course,
the bodies may be supposed to exhibit altogether different elastic
powers, and laws of expansion.

chemically with the molecule of oxalic acid ; and thus completes the molecule of the acid, as it actually exists in the crystalline form.

Such are the views we have been induced to take of the nature of chemical combination: whether right or wrong, they have the merit of being exceedingly simple, and consistent with themselves, thoughout ; which can hardly be said of any others, with which we are acquainted. Indeed much reflection upon the subject, for many years past, has satisfied us, that chemical combinations can be rationally explained only, in some such manner as we have supposed. Any lengthened argument, however, on the laws of chemical combination here, would be quite out of place ; we shall therefore confine ourselves to the following observations.

First. The above view of the molecular constitution of bodies, naturally suggests the question : do the sub-molecules, which we suppose to unite together cohesively, and form the self-repulsive molecule, of oxygen and hydrogen, for instance, possess the same properties, as those of oxygen and hydrogen ? or do they possess different properties ? These questions, in most instances, cannot, in the present state of our knowledge, be satisfactorily answered ; though there is every reason to believe that the properties, both of the sub-molecule, and of the super-molecule, generally differ from those of the molecule itself; but that the differences are rather of a specific, than of a generic character.* Thus chemists have shown that different volumes of the same gaseous body, termed carburetted hydrogen, combine together, and form various compounds: we have, for example, a gas, one volume of which contains two volumes of carburetted hydrogen; another, one volume of which contains three, and another four, of the same gaseous body. Now the sensible properties of all these compounds, though resembling each other in some respects, are yet specifically different: and as they are all composed of the same gaseous body in different proportions; these differences must be considered rather as the result of cohesive, than of chemical, union. Thus the supposition, that both the sub-molecules, and the super-molecules, of bodies may possess properties different from one another, and from the standard molecule, is rendered exceedingly probable, by the above facts; and if our space admitted, it would not perhaps be difficult to bring forward other facts of the same kind. This however would be foreign to our purpose; and

---

* What we term the *sensible* properties of bodies are, of course in all instances, the result of a great number of molecules acting together at the same time; hence below a certain point, mere difference of numbers may be supposed to produce a change in sensible properties, not only in degree, but in kind: of the sensible properties of a single molecule we can form no conception.

we shall only remark, that a great many curious circumstances, at present but very imperfectly understood, evidently appear to be referrible to a similar principle.

Secondly. Although we have thus rendered it probable, that the molecules of bodies considered at present as elementary, are immediately compounded of many others, more or less resembling them; yet it is obvious, that there must be a point at which these, and other elements, exist in a primary or ultimate form; and beyond which, if they can be supposed to be subdivided, they must become something altogether different. In this respect, therefore, the views we have advanced, accord generally with those at present entertained; and the only point in which they differ, is in supposing, that the self-repulsive molecule, as it exists in the gaseous form, does not represent the ultimate molecule; but is composed of many of them. With respect to the nature of the ultimate sub-molecules of those bodies, which we consider at present as elements, as, for instance, of oxygen; they may naturally be supposed to possess the most intense properties, or polarities. Indeed, such sub-molecules may be imagined to resemble in some degree, the imponderable matters, heat, &c., not only by their extreme tenuity, but in other characters also; and this very intensity of property and character may be reasonably considered as one, if not the principal reason, why they are incapable of existing in a detached form. Lastly, are not these ultimate and refined forms of matter extensively employed in many of the operations of nature ; and particularly in many of the processes of organization?

Thirdly. By supposing that these laws of combination are not confined to elementary bodies, but extend to all others throughout nature ; and that bodies, however complicated they may be, always act as simple molecules; and always combine with reference to their volume in the gaseous state ; we are enabled in some degree, to explain that endless variety of property, and condition, which we see around us. For no sooner is a new compound molecule formed by an assemblage of similar molecules, than it may be supposed to be capable of combining with other molecules *chemically*, and of thus entering into a long and novel series of combinations : while these combinations again in their turn, may be imagined to lead to others ; and so on, till the variety becomes extreme. Indeed, were not such combinations limited by the very nature of things themselves, no two substances would probably possess the same properties. As it is, most of these compounds are incapable of separate existence : thus the compound super-molecules of water in the crystal of oxalic acid before referred to, are incapable of separate existence : if they could exist separately, would they assume the form of water?

Fourthly. It would not be difficult, though not very safe, or prudent, in the present state of our knowledge, to speculate on the crystalline forms assumed by different bodies, with reference to the principles we have advanced. We shall therefore not touch upon this part of the subject, further than by observing, that the cohesive force, though supposed to possess some peculiarity, as existing *among the molecules of different bodies*, is nevertheless essentially but of *one* kind. When therefore, the molecules of different bodies are of the same size, (or rather of the same weight,) they may be naturally supposed capable of associating themselves into the same form ; and if they happen to be mixed together, they may even enter indiscriminately into the same crystal. Hence arises what has been termed the *isomorphism* of bodies ; while if there be a near approximation, but not an exact coincidence in the above relations, they may, upon the same principles, be supposed to give origin to *plesiomorphism ;* that is to say, to a near approach to a similarity of form.

Fifthly. With respect to the nature of the circumstances which determine the characters, and modes of existence of bodies, we know very little. We are almost equally ignorant also, of the nature of the causes which determine the cohesion of the molecules of bodies, into the crystalline form. A variety of arguments might, however, be brought forward, which appear to show, that the size, and shape, of the molecules, have a great deal to do with crystallization ; certainly, at least, the molecules must be supposed to have a size, and shape, somehow or other adapted for the modes in which they are arranged ; otherwise they could not be capable of such an arrangement. The cause of this similarity of size, and shape, is unknown ; but it most probably depends upon the *similarity of weight, (Isobarism)*, of the molecule ; that is to say, upon the *relation or identity of the absolute quantity of matter which the molecule contains ;* which relation, as far as we can perceive, is not only the sole circumstance common to the molecules of different bodies ; but that which, of all others, is the most likely to produce identity in the size, and shape, of these molecules.

Sixthly. When the molecules of bodies in solution do not happen to possess the requisite size, and shape for cohesion, there is, from the phenomena, reason to believe, that they occasionally possess the power, as it were, of making up the necessary form, by attaching to themselves the molecules of other bodies. Now, bodies so attached may be considered as acting a sort of complementary part ; that is to say, they may be supposed to *complete* the size, or figure, of the molecule; so as to adapt it for combining in a certain manner. Thus, the water of crystallization, (and perhaps occasionally other matters), appears in the greater number of instances to perform an office of this kind; and to be, in fact, strictly complementary to that particular size, and figure, of the molecules, which may be supposed to be requisite, for enabling them, not only to combine the more readily with each other; but at the same time, to form a symmetrical solid, or *crystal.**

One or two other circumstances, connected with this part of our subject, will be better understood after we have considered, a little more in detail, the combinations of bodies with reference to their *weights;* and the absolute quantity of matter, which they contain. To this most interesting inquiry, therefore, we shall in the next place proceed, confining ourselves, however, as

* There is every reason to believe, that one variety of isomorphism is effected on the principles here stated; and that the molecules of *different* substances, by attracting to themselves *different* quantities of water, or of other matters, may ultimately make up compound molecules, similar to those of the bodies with which they may happen to be mixed; and may thus enter indiscriminately with these bodies, into the crystalline form. Such a state of things is calculated to baffle the mere chemist, however expert; though it is probable, that if carefully examined and understood; an intermixture of this kind might be detected, by the optical properties of the crystal.

before, principally to the elements of water, hydrogen and oxygen.

It has been found by experiment, that the same volumes of different bodies in the gaseous state, have very different weights. Thus, for instance, a volume of oxygen weighs sixteen times as much, as the same volume of hydrogen. Hence, as the number of self-repulsive molecules in each of these gases, is presumed to be the same; the weight of the self-repulsive molecule of oxygen must of course be sixteen times greater, than that of hydrogen; and GENERALLY, *the weights of the self-repulsive molecules of all bodies, will be as the specific gravities of these bodies in the gaseous state; or will bear certain simple relations to these specific gravities.* This relation in *weight* among the molecules of bodies, constitutes the basis of what is called, the *Atomic theory,* proposed, some years ago by Dr. Dalton; who established the most important fact, that bodies do not, as formerly supposed, combine at random, but in definite proportions by weight; and if the preceding doctrines be well founded, it is evident they cannot combine otherwise.*

As however water is composed of one volume of

* The reader is referred to " An Introduction to the Atomic Theory," recently published by Dr. Daubeny, Professor of Chemistry, at Oxford; for an interesting and able inquiry into the principles of this theory.

oxygen, united with *two* volumes of hydrogen, the relative weights of the hydrogen and oxygen in water will be, not as 1 to 16, but as 1 to 8 only; while the weight of the self-repulsive molecule of steam, will be 9. Hence, as one, or the other, of the elements of water, is usually made the basis of the atomic numbers, this difference between the volumes and the combining weights of its elements, has produced considerable confusion; and has given rise to much needless discussion. As a mere matter of convenience, it is certainly preferable to consider the two volumes of hydrogen, as one *atom*, (to use the language of Dr. Dalton); in which case, oxygen will be 8, and water 9; but a strictly philosophical arrangement, supposing the principles we have advanced be well founded, would require, that *the volume in all instances should be made the molecular unit;* in which case, the relative weights of the self-repulsive molecules of hydrogen and oxygen, as above mentioned, will be as 1, to 16.

In this country, two volumes of hydrogen, as we have said, are usually considered as *one atom*, or unity, in which case, oxygen is 8; but some have chosen instead of hydrogen, to make oxygen unity, or 10; in which case, hydrogen, of course, will be the one-eighth of 1, or of 10; that is to say, ·125 or 1·25; and .water, instead of 9, will be 1·125, or 11·25. It matters not which of these series of numbers, or whether any other be employed, so that the same relative proportions be observed among them; but the first series is that most generally adopted, and is upon the whole the most convenient. In the above manner, the *atomic weights*, as they are termed, of all bodies capable of assuming the gaseous form, can be easily obtained; but in those bodies that do not assume the gaseous form in their simple state, but in some state of combination only, we are obliged to deduce the weight of the primary molecule, from that of the compound. Thus carbon, in its elementary state, is incapable of assuming the gaseous form; but combined with oxygen, it forms carbonic acid gas; one volume of which, weighs 22 times as much, as our standard two volumes of hydrogen. Now it has been found by other experiments, that of these 22 parts, 16 are oxygen. The remaining 6 parts must, therefore, be carbon; and accordingly 6, is the number upon our scale representing carbon; and the proportion, with reference to which, this body always enters into composition. In the case of bodies, as for instance lime, which are incapable of assuming a gaseous form, either alone, or in combination, we are obliged to trust solely to analysis; thus common marble, or carbonate of lime, as it is termed by chemists, is found to be composed of 22 parts of carbonic acid, and 28 parts of lime; 28 therefore represents upon our scale, the atomic weight of lime; and so of all others.

It may be observed, that we have spoken as if the atomic weights of bodies, were related to one another by multiple; that is to say, were all multiples of some common unit. Now this opinion has been maintained by some; while it has been denied by others; who admitting that multiples in weight are necessary to the union of the *same* body, both *chemically* and *cohesively;* will not admit, that they are necessary to the union of *different* bodies. The matter is one that in the present imperfect state of chemistry, can hardly be determined by experiment; for what with the difficulty, or rather impossibility, of procuring bodies in a perfectly isolated form, and the unavoidable imperfections of all chemical processes; we can scarcely hope to approach, within the necessary limits of precision. If the above views of molecular relations, however, be well founded, it seems almost impossible to arrive at any other conclusion, than that the combining weights of all bodies are intimately related by multiple; though to enter further upon the subject here, would be quite foreign to our present purpose.*

* For the sake of those who are interested in such matters, one or two of the leading arguments may be briefly stated. We have rendered it probable, that when two or more molecules of

Lastly, it may be remarked, that the numbers at present conventionally employed by chemists, to represent what have been called the *atomic weights* of bodies, are so convenient, that they will not readily, nor indeed ought lightly, to be set aside ; though there is reason to believe that many of them require revision, and are destined to undergo material alterations, even as the subject is at present understood.   If the views however which we have advanced be correct, these numbers certainly do not represent nature : for as we have already stated, a strictly philosophical arrangement can be rationally founded only, upon the volumes of bodies in the gaseous

the same body combine *cohesively*, they form a compound, which though having properties in some degree allied to those of the original molecule, nevertheless usually possesses a specific difference ; that is to say, the *chemical* polarities of the compound molecule as modified by the union, will be different from those of the simple molecule.   But a body possessing a *specific difference*, may be supposed to be a *new* body; and thus capable of combining in future, *not* cohesively, but *chemically* with our original molecule.   Now in such a case, it is evident that the weight of the original molecule, and that of the new compound molecule, *must* have a certain relation to one another, by multiple. If our space admitted, it would not, we believe, be difficult to point out instances of such combination among chemical phenomena ; but we shall merely observe, that many of the substances, at present considered as elementary, appear to be constituted upon the above principles, from some common molecule, of a still more elementary character.   Moreover this law seems to hold universally throughout nature; and those substances related to the same molecule, in general constitute a natural group or family, having certain properties in common.

P.   M

state; in which state, some common volume in all instances, should be considered as the molecular unity. Now, as in most instances, this molecular unity seems capable of sub-division; of course the number made to repre-sent it, can hardly ever be supposed to be a *prime* number. Hence, as combining molecules of bodies exist both below and above the mole-cular unity, they may often, (perhaps always), be represented by a series. Thus suppose 9, to represent the molecular unity, or volume, of water; and that this be subdivided into three (which it is at least, and probably into a much greater number), the molecular combinations of water may be represented by the series, 3, 6, 9, 12, 15, 18, &c. We mean to say, the molecules of water, as they actually enter into combination in different bodies, may be supposed to be re-presented by these numbers; while, by way of distinguishing the different molecules, those below 9, may be designated generally *sub*-mole-cules; and those above 9, *super*-molecules; and the molecular unity itself may be simply called the *molecule;* or in the gaseous state, the *self-repulsive molecule;* distinctions, which for the sake of convenience, we have adhered to throughout these remarks, and which we have thought it thus necessary to explain.*

* The above terms are to be considered as a temporary expe-dient only. If these views be established, it will not perhaps be

SECTION IV.

Recapitulation of the last Section. General Reflections on the Subjects treated of in the preceding Chapters.

THE subjects considered in the last section of the present chapter, may be viewed as a continuation of what has engaged our attention, in those that have preceded ; and the principal circumstances detailed, may be thus recapitulated.

1. All perfectly gaseous bodies combine with reference to their volume ; that is to say, any volume or bulk of a gas, always combines with an equal volume or bulk of the same, or of another gas : or with a volume, having some simple relation to its own volume, as half, or twice as much ; &c. and not with any intervening fractional part of a volume.

2. The same volume of different gaseous bodies has very different *weights :* hence on the supposition formerly advanced, that all perfectly gaseous bodies under the same pressure and tempe-

difficult to devise hereafter both a notation and a nomenclature founded upon them. At present such an attempt would be ridiculous.

rature, contain an equal number of self-repulsive molecules, the molecules of different gaseous bodies must also have different weights ; which weights will be as the specific gravities of the gases; and may be represented by numbers proportional to these specific gravities.

3. From the above relation between the volumes and the weights of bodies in the gaseous state, it follows, that *all bodies must combine with reference to their weights;* that is to say, that the same weight of the same body, (or half or twice as much, &c.) must always combine with the same weight, (or half or twice as much, &c.) not only of the same, but of every other body.

4. The numbers representing the relations among the specific gravities of bodies in the gaseous state, are called the *molecular*, or *atomic weights*, of the different bodies.

Such is the foundation of what is usually called the *Atomic Theory ;* the principles of which theory, are generally acknowledged to regulate chemical combinations.

We shall now conclude the present Treatise on chemistry, with a few remarks, more especially relating to the object of these volumes. And here it may be observed, once for all, that throughout the preceding pages, as well as in what follows, we have endeavoured to state each argument as distinctly as possible, without encumbering it too much with details—in short, to illustrate principles, rather than to enumerate particulars. When the principles of a cumulative argument are understood, the details are readily supplied by the reader.

First. On taking a general and collective review of the facts brought forward in the preceding chapters, the circumstances calculated to strike our attention in the first place, are the wonderful coincidence between the priority of existence, and the universal prevalence, of the primordial agents and elements of nature, on the one hand ; and on the other, the beautiful adaptation of the agents and elements of a later, and more subordinate character to these primordial principles ; so that, when the whole are taken together, they constitute one harmonious and connected series, in which all the various parts are mutually adapted, and dependent. In the following chapters, we shall have occasion to notice many of the more important of these subordinate arrangements ; at present, we shall chiefly confine ourselves to a general review of what has been already stated.

We are told by the inspired historian, that after matter had been created, and endowed with motion, the next Almighty fiat was, " let there be light;" and if we suppose this fiat, to have included the other imponderable forms of matter, heat, &c. ; how entirely do the whole phenomena of nature accord with the sacred narrative? Light, and probably its attendant heat, are the most generally diffused and universal of all the subordinate agencies; so much so, that they are not confined to our globe, or even system, but extend throughout the universe. Their laws and influences, therefore, seem to be as general, and as necessary to the present order of things, as those of gravitation itself. The priority of existence also, of light and of heat, is self evident; for until they existed, nothing else, as we are acquainted with things, could have had existence. Now all subsequent creations have been made with the most exact regard to the influences of these prior agencies. The globe, for example, which we inhabit, is placed at a certain distance from the sun, the great centre of our system, and of light, and of heat; and where of course, according to the laws which light and heat obey, these agencies must act with a certain intensity. Hence it was necessary that the materials of our globe should have a certain degree of fixity; otherwise they could not exist. If, indeed, there had been no ulterior views, with respect to the destination of our globe; all that would have been requisite, would have been, to have made it sufficiently firm to move through space; and for this purpose the more homogeneous, and compact its composition had been, the better. But what are the facts? Our globe, though stable; so far from being homogeneous, is composed of a variety of substances, all differing from each other in their properties; some being solid, some fluid, some aeriform, under the common circumstances in which they have been placed; and all beautifully adapted, both by their physical and chemical properties, to the purposes they fulfil in nature; nay, what is more, to the purposes they were *designed* to fulfil in nature; for on no other supposition, would their properties be intelligible.

Thus water, *within very narrow limits of temperature*, is a solid, or a liquid, or a gas; and yet these *very narrow limits of temperature*, neither more nor less, are precisely those, which exist upon the surface of our globe; where they are the natural, and the necessary results of its situation in the universe; and of the general laws, which govern the distribution of light and heat. Had the properties of this body been other than what they are; or had the general temperature of our globe been different; water would have existed altogether in the solid, or in the gaseous state; and its most important properties would have been unknown. Hence, it seems almost impossible to arrive at any other conclusion, than that the temperature of the earth, and the properties of the water on its surface, have been mutually adjusted to each other. And further, since the temperature of the earth, as just stated, is the natural result of the general laws which govern the distribution of heat and of light; the inference must be, that the properties of the water, as the subordinate and later principle, have, at an after period, been adjusted to the prior temperature of the earth.

If we do not admit of this adjustment, we must suppose that the whole has been the result of chance, or of some other unintelligent principle ; and if water had been the only principle in which such adaptations were apparent, the supposition of chance might, perhaps, be received ; at least it would be difficult to prove the contrary. But when we see similar happy adjustments in every object around us,—in the different elements of the air we breathe, in the soil we tread upon, in all the varieties of rocks composing the solid crust of our globe; not one of which could have been more happily contrived for the purposes they fulfil, nor indeed be scarcely conceived to exist otherwise than as they are, without destruction to the whole of the present arrangements—when we see all these things, and duly reflect on them ; it becomes absolutely impossible to suppose, that so much happy adjustment; so much apparent intelligence ; so much, in short, of what the veriest sceptic, under other circumstances, would have allowed to be evidences of design, can be evidences of any thing else than design ;

or have resulted from any unintelligent cause whatever. Hence we are driven irresistibly to the only rational conclusions which the premises appear to admit of, viz. ; that all these happy adjustments, and adaptations, which we see in nature, are really and truly what they appear to be,—so many evidences of design ; and, consequently, that the whole have sprung from the will of an intelligent, and omnipotent Creator.

The above inferences are deducible from the plain and obvious arrangements of nature, which every one can readily understand : but when speaking of elementary bodies, we remarked, that in a variety of instances, their object and use were unknown to us ; and before we quit this part of our enquiry, it may not be out of place, to consider briefly these difficult points.

When we see adjustments so wonderful, and such wisdom displayed in those parts of creation which are intelligible to us ; we cannot imagine that the Being who made them all, would act otherwise, than with wisdom. Hence, what we do not understand, or what may appear incongruous to us, we naturally and properly refer to our own ignorance. The phenomena of chemistry are so extraordinary, and often so unexpected, that little in general can be predicated of them, beyond what is actually known. The most experienced chemist, therefore, as compared with the Great Chemist of nature, is immeasurably deficient; and can only contemplate His wonderful operations with astonishment and awe, and own them unapproachable. Who then can tell what design is latent, under *apparent* incongruities? What elaborate contrivances, and adaptations, may have been requisite to have produced water, or carbon, or any other essential principle, out of the materials, and in conformity to the laws, by means of which the Great Author of nature chose to operate? Who can tell that the minor evil, may not have been essential to the existence of the greater good? That the poisonous metals, for instance, are not, as it were, the refuse of the great chemical processes, by which the more important and essential principles of nature have been eliminated? That these poisonous principles have not been left with such subdued properties, as scarcely to interfere with His great design,—not because they could not have been prevented—not because they could not have been removed—but on purpose, and designedly, to display his power?

Secondly. If we pursue the subject a step further, and inquire into the means by which all the beautiful adaptations we have been considering, are effected, we shall find, that they principally depend upon a certain due adjustment, to each other, of the *qualities,* and *quantities* of the different substances; and more especially, of the different elementary principles, of which our globe is composed. These adjustments are so universal, and so varied in their character ; that to enumerate them all, would be little else than to enumerate all the objects in nature ; we shall therefore content ourselves with a few of the most familiar of each kind.

In the first place, with respect to the adjustment of *quality*. Let us consider for a moment, and by way of illustration, what would happen, if the qualities of water, or of air, were to undergo a change : were, for example, the important fluid water to become sour or sweet ; or heavier or lighter ; or indeed any thing but what it is : or were the air of the atmosphere to acquire odour or colour ; or to become opake : by either of such changes, slight as they appear, the whole of the present economy of nature would be deranged. Again, if the qualities of the acid, existing in the common salt of the ocean, were to become so modified, as to quit the alkali with which it is at present associated, and combine with the limestone composing our rocks ; while the carbonic acid, thus set free, was diffused through the atmosphere : in such a case, a large part of the solid crust of our globe would rapidly disappear, and, becoming dissolved in the waters of the ocean, would totally unfit them for their present purposes ; while the liberated carbonic acid, would instantly prove fatal to animal life. These would be the con-

sequences of changes so trifling in the qualities
of a few substances only; nor is it possible,
scarcely, to conceive any other change, that
would not be attended with similar results.

In the next place, the importance of the adjustments of *quantity*, is equally striking. Let
us, for instance, conceive what would happen
from the simple inversion of the quantities of
dry land, and of sea, as they now exist: in such
a case, there would not be enough of water to
preserve the surface of the land in a moist state;
the greater part would thus be in the situation
of the deserts of Africa; and totally unfitted for
the habitation of organized beings. When
speaking of the elements of water, we alluded
to the happy adjustment of the quantities of
oxygen and of hydrogen in the world; and to
the consequences which would have ensued, if
hydrogen, instead of oxygen, had predominated.
The same remarks apply to almost every other
element; for example, had the proportions of
the chlorine, and of the soda in common salt;
or of the carbonic acid, and of the lime, in our
marbles, been different from what they are; the
one or the other of the ingredients must have
been in excess; and the present order of things
could never have existed. Again, were gold
suddenly to become as abundant as iron, and
iron as rare as gold; were the carbon existing in
the present useful form of fossil coals, to assume

the crystallized form, and become diamonds; the whole order of nature would be subverted, and the whole of the present arrangements be involved in ruin. Those who deny the argument of design, of course consider such suppositions as these absurd; and if carried too far, they doubtless, under any circumstances, lose much of their effect; but admitting the argument of design, the judicious application of such suppositions, is well calculated to place the advantages, and effects of certain arrangements in a more striking point of view, than can be obtained by any other means. More especially, such suppositions, by showing the wonderful adaptations of subsequent creations, to prior existences, are admirably calculated to illustrate their fitness, and consequently, the *apparent* design, displayed in the formation of the prior existences; and thus to show, that these prior existences must have been created with reference to ulterior purposes.

The argument of prior arrangements, and of the subsequent adaptation of other creations to these arrangements, is one of such interest; its consequences, also, are so important; that perhaps it may not be deemed irrelevant, if, for further illustration, we recapitulate the argument in a condensed form. For this purpose, we shall select the obvious and familiar relation of plants and animals to water and air.

The prior existence of water and air, as compared with that of plants and animals, is established by the fact, that water and air *can* exist without plants and animals ; but that plants and animals *cannot* exist without water and air. Hence, as water and air must have existed, with all their present properties, before plants and animals were created ; the question naturally arises, how water and air came to be endowed with their present properties? *We* suppose that water and air were created with their present properties, with reference to the future existence of plants and animals ; and on this supposition the whole becomes intelligible. Further, that this is the true explanation, and that water and air have not obtained their present properties, by chance, or accident, is rendered still more probable by the following considerations. We have said that water and air can exist without plants and animals : now as far as we know, water and air might have existed *for ever* without plants and animals ; at least the contrary cannot be proved, or even rendered probable. Moreover, plants and animals, as involving new principles of a higher order (those of life), never could, by any law of nature, necessary or probable, have resulted from an inferior agency. Hence, there is no necessary relation of cause and effect, between the prior existence of water and air, and the subsequent existence of plants and animals, as some seem to have supposed. Hence too it follows irresistibly, that plants and animals have been created, and their properties adapted to those of water and of air, at some subsequent period, and by some external and superior agent. But the agent that could thus create plants and animals, could surely have created the water and air likewise; nay, *must* have created them; for, as the prior and subsequent creations taken together, evidently form but different parts of one and the same general design, the whole design must have been the work of one and the same intelligent Agent.

It yet remains to draw the attention of the reader to another circumstance, connected with these adjustments in quality and quantity, viz. the *double* adjustment. Of the causes of the qualities of bodies we know but little, and that little is founded solely on experience. We see that these qualities are admirably fitted for their apparent purposes; and hence, as they might have been different, we arrive at the probable conclusion, that they have been so fitted by design. The collocation of quantities and numbers, exactly where they have been required, adds much to the probability of this conclusion; as such a collocation could hardly have been other than the act of an intelligent Being. But the *double adjustment in quality and quantity, of the same thing at the same time,* adds almost infinitely to the weight of evidence; and indeed furnishes a proof in favour of design, and of its consequences, which amounts to all but actual demonstration.

Thirdly. There is another point of view in which we may consider what has been stated, and by which we shall at the same time,. be brought a step nearer to the existing order of things. Amidst all that endless diversity of property, and all the changes constantly going on in the world around us, we cannot avoid being struck with the general tendency of the whole, to a *state of repose*, or *equilibrium*. Moreover, this tendency to equilibrium is not confined to the ponderable elements, but prevails also, in the same remarkable degree, among the imponderable agencies, heat and light; which, as we have seen, cannot be any where long retained in a state of excess, on account of their natural disposition to acquire a certain state of equilibrium; depending generally upon the place of the earth in the solar system. Now, the formation of this state of equilibrium, and its preservation, may be considered as the results of those wonderful adjustments among the qualities and quantities of bodies above alluded to;—the qualities being such as to neutralize each other's activity; while the quantities are so apportioned, as to leave one or two only, predominant.

The preceding is a general view of the subject. But it is to be observed, that the state of equilibrium here described is not absolutely *fixed;* as such an unyielding condition would be not less incompatible with the present order of things, than a condition of unlimited change. The whole are so adjusted therefore, that slight deviations, or oscillations about the neutral point of rest or equilibrium, take place, and are even necessary, as the world is at present constituted; though these changes are bounded within very narrow limits, and greater deviations would instantly prove fatal to the whole. If we enquire into the principles upon which these slight deviations take place, and are regulated; we shall find still further reason to admire the wonderful arrangements displayed. When speaking of the elements of water, we observed how much the stability of nature depended on the proportions of the elements of this fluid; and that one of its elements, oxygen, existed in excess, and in a free state, in the air. Now, it is to the agency of this oxygen in a free state, and to the annual and diurnal motions of the earth, that most of the minor operations going on around us are to be referred. The universal presence, and peculiar properties, of oxygen are such, as to interfere more or less, with every thing; while the motions of the earth, keep every thing in a constant state of activity and change. Yet, the general tendency of the whole, as before ob-

P.   N

served, is towards a state of equilibrium; and the principles upon which this tendency operates, are very intelligible. Thus, all bodies *below* the neutral point of rest, if we may be allowed the expression ; that is to say, all bodies of a marked elementary character, have a tendency to combine with each other *synthetically;* while *beyond* the neutral point, bodies have very little tendency to combine further; and if by intention on the part of the operator, or from any other cause, they be so made to combine ; when left to their own operations, they speedily revert, or oscillate back, to the point of equilibrium.

Such are the means by which the state of equilibrium we are considering has been produced, and by which it is still preserved; nor is it possible to reflect upon the subject for a moment, without arriving at the conclusion; that this state of equilibrium possesses all the characters of a prior arrangement, to which organized beings have been subsequently adapted. We are thus led, in the next place, to make a few remarks upon the subsequent adaptation of organized beings, to the pre-established equilibrium of nature.

The present races of organized beings, are, in all instances continued, only by the process of generation ; and if they were annihilated, there are no natural operations going on in the world, which can lead us to believe, that by any law of

nature, such organized beings could be repro-
duced. That is to say, we cannot conceive that
hydrogen, carbon, oxygen, and azote, with heat
and light, &c. from what we know of their properties, would ever be able, of their own accord,
so to combine as to form a plant or an animal.
Hence, when plants and animals were first produced, it is evident that there must have been a
power or agent in operation, which has long
since discontinued so to operate; and that this
power or agent not only created plants and
animals; but at the same time imparted to them
a capability of perpetuating their existence, for
a period, at least, commensurate with that state
of equilibrium in which they have been placed.
Now, whether we consider the power or agent
who accomplished all these things, to have been
the Deity himself operating immediately, which
is most probable; or whether we consider with
some, that He operated by delegated agencies
and laws, the result is the same as far as our
argument is concerned; the object of which
argument is to show, that the present races of
organized beings are, somehow or other, influenced by the same general laws, which appear
to regulate inorganic matters. That is to say;
organized beings at the present time, are at least
as fixed and permanent in their nature, as the
state of equilibrium in which they have been
placed; and consequently, no new plants or new

animals, can, as the world now exists, be imagined to be produced, without a new and specific
act of creation; or at least, without an entire
change in the standard of equilibrium.

We have alluded to the commencement of the
present order of things, and to a possible state
of change in the condition of equilibrium: perhaps, it may not be amiss to make a few further
remarks upon these points. That the present
order of things, most certainly has had a beginning; and as certainly, will come to an end;
we cannot doubt; the questions are, when was
this beginning; when will be this end? Of the
end, of course, we can know nothing: the beginning is less obscure; and there are indelible
impressions left upon the materials and structure
of our globe, which throw no ordinary light upon
this question. The consideration of the changes
which our earth has undergone, however, belongs to another department: we shall only
observe, that these changes appear to be of two
distinct orders; which have alternated with one
another in succession. The first of these orders
of changes, seems to have been of a slow and
gradual kind; and such as might be supposed
to take place, during a state of things, more or
less like the present, and existing for a considerable period. The changes of the second
order, on the contrary, have evidently been violent, sudden, and disruptive; of comparatively

short duration, and differing exceedingly in degree, and in extent. In general, they appear to have operated from within; but whether altogether from internal, or from external influences, is unknown to us. Now, it is remarkable, that these successive alternations seem each time to have changed the standard of equilibrium; and that during the state of comparative quietude, or the interval of equilibrium between the convulsions, organized beings have existed, adapted to the exigencies of that particular state of equilibrium; and which beings must have been successively created: moreover, the later creations gradually approach to those at present in existence. Hence, not only does the change in the standard of organization, seem to have been simultaneous with the change in the state of equilibrium; but both appear to have been progressively raised after each convulsion. Finally, the last general catastrophe of the disruptive order was evidently a deluge.* Such are the conclusions, which geologists have deduced from a careful survey of that part of the crust of the

* If we judge from what is going on around us in nature, and from the little tendency there appears to be in things, at present, to combine into new forms; we must be almost led to the conclusion, that the developement of new elements, as well as of new agents, is necessary to produce new and specific arrangements. May we not then infer, that during those periodic convulsions alluded to in the text, new elements have been developed, or old ones decomposed into others of a higher, and more

earth to which they have access; and these conclusions are of the most important kind. In particular, by demonstrating the existence of successive adaptations, to successive and different states of equilibrium; they place the argument of design in a new light, and add, in no small degree, to its force. This part of the subject, however, belongs to the geologist, to whom, for the present, we shall leave it.

Fourthly. The argument of design, as connected with the subject of equilibrium above treated of, may be considered yet in another point of view. In this state of equilibrium we have observed, that the properties of bodies, as they actually exist around us, are all so subdued and passive in their character, that no one predominates over, or excludes the others. Now, when we reflect that almost all these bodies are *compounds;* and when we compare the properties of these compounds, with the properties of the elements composing them; it is impossible not to infer, that the properties of the compounds, rather than those of the elements, were,

elementary kind; and that in virtue of the general laws in operation, these new elements have subsequently combined to form series of new arrangements? Of course, this supposition is intended to apply only to the *means* adopted by the Deity to effect his purpose. The formation, and selection of these new elements, must in all instances, be supposed to result immediately from His will and agency.

at their origin, the objects contemplated. That is to say; in order that the compounds might be perfect, the elements calculated to produce them, were created essentially such, as these compounds might require; without reference to the secondary properties of the elements themselves; which were left to be determined, as the more general laws of matter might decide. For instance, the hydrogen in water, and the chlorine and sodium in common salt, not being, in their simple state, required in the economy of nature; the properties of these elements have not been made compatible with organic existence; and the whole attention, (if such a term may be applied to the operations of the Deity,) has been directed to the properties of the compounds, water, and salt. Thus, on the one hand, where required, we have the most striking adaptation of property; while on the other, where *not* required, this adaptation of property has *not* been attended to: nor is this true of water and salt only, but of almost every other compound in nature. Nay, what is more, the incongruities of the whole system have, with the most consummate skill, been thrown, as it were, among those properties *not* required. Hence, the arrangements of nature viewed in this light, not only exhibit novel evidences, but some of the most striking evidences of design, which we possess.

The subject of the incongruous properties of bodies, is one of great interest. We have seen that many of the elementary principles are poisonous; and that almost all of them, if liberated from their affinities, and sent abroad in the world, like so many demons let loose, would instantly bring destruction upon the whole fabric. Now, why should such incompatible properties be necessary to the properties of the compounds? Why, for instance, should the incombustible fluid water, contain one of the most combustible principles in nature? Or the mild and innocuous common salt, be composed of two elements, which, in their separate state, would instantly destroy life? Why, we repeat, are these deleterious properties of the elements, necessary to the wholesome condition of the compound? What part do they perform; or what property do they represent, or modify? These are questions utterly beyond our comprehension; and are likely always to remain so. That these incompatible properties of the elements, however, do, in some way, contribute to the perfection of the compounds, we cannot doubt; and the only grounds, upon which such incompatibility seems to admit of explanation, is; that it results necessarily from those limitations, which the Deity has thought proper to prescribe to his power; and to which He always most rigidly adheres. Moreover, be the reason what it may;

it is evident that these arrangements, so immediately calculated to lead to practical difficulties,
have been the result of *choice*. For we cannot
but believe that an omnipotent Creator, if He
had so willed, could have made the elements
innocuous, as well as the compounds; nay, to
our limited understanding, this would have been
the easiest, and most natural, mode of proceeding. Why then did He choose the apparently
more difficult course? Why, to use the language of Paley, but " that He might let in, and
thereby exhibit, demonstrations of his wisdom."
Throughout nature, the exigences and incongruities necessarily arising from the arrangements we have been considering, have given
occasion for the display of the most astonishing
wisdom and power. And instead of that jarring
and clashing, which might have been expected
from so many conflicting elements, the qualities
and quantites of these elements have, upon the
whole, been so wonderfully adjusted to each
other, that they neutralize and balance each
other's evils; and the general result has been,
that all have finally settled down together, into
that harmonious state of equilibrium, before
alluded to, so admirably adapted for the existence of organic life.

Fifthly. We have hitherto confined our attention to general principles, and arrangements; but
the commonest chemical process may be made

to furnish us, with some striking proof of the
omnipotence of the great Creator. Let us, for
example, consider what happens in a simple and
familiar instance of chemical decomposition; as
when a solution of lunar caustic, (nitrate of
silver,) is added to a solution of common salt.
In this case, the chlorine of the salt combines
with the silver, and produces a curdy precipitate,
which falls to the bottom; while the nitric acid
combines with the soda, and forms a soluble salt,
which remains in solution. Now, we showed in
a former chapter, that the minutest fragment of
matter appreciable by our senses, consists of innumerable molecules. If therefore we suppose a
small quantity, as an ounce, of the lunar caustic,
and a proportionate quantity of common salt, to
be mixed together; what countless myriads of
molecules, in a portion of time literally inappreciable, must have sought out, and combined,
each with its fellow, in this simple process! The
human mind absolutely recoils from the contemplation of objects so completely beyond its
powers; for the utmost that we can imagine,
must fall almost infinitely short of the reality.
Were we, for illustration, to conceive all the
human beings at present in existence, to be collected together into one vast array, and to be all
dressed exactly alike, and to perform the same
military manœuvre at the same moment; we
should be probably as far short of the actual

numbers of similar molecules, each manœuvring exactly alike, in the above simple experiment, as a single company, falls short of our congregated army! Again, to take another familiar illustration, as the working of a common steam engine; we are assured, that in this simple operation, there are more self-repulsive molecules of water always constantly engaged, and conspiring to the same end, than there are quadrupeds in existence upon the whole surface of the globe! The above are designed to illustrate the principles of the argument only: the argument itself, like all the preceding, is strictly cumulative; and applies, more or less, to every operation in nature.

Such is a summary sketch of the wonders developed by chemistry; and what an idea do they convey to us of the wisdom, and of the power of Him, who contrived and made the whole! Of the *capacity* of that eternal Mind; who, while He directs the universe; at the same time, takes cognizance, and regulates the movements, of every individual atom in it! To whom, the inmost nature, and end, and object, of every part are familiar; of whose comprehensive designs, the whole forms but a single link; the antecedent and the consequent to which, are merged alike, in infinity!

BOOK II.

OF METEOROLOGY:

COMPREHENDING A GENERAL SKETCH OF THE CON-
STITUTION OF THE GLOBE ; AND OF THE DISTRI-
BUTION AND MUTUAL INFLUENCE OF THE AGENTS
AND ELEMENTS OF CHEMISTRY IN THE ECONOMY
OF NATURE.

In the First Book, we have endeavoured to
convey some notion of the " limits which the
Deity has been pleased to prescribe to his own
power ;" or in other words, to briefly describe
the properties of the different subordinate agents
and elements of our globe ; and their laws of
operation. We come now, to consider a little
more closely, the general distribution of these
agents and elements ; and the principles upon
which this distribution is regulated ; so as to
produce all the wonderful results, which we see
constantly going on around us, in nature.

In the present state of the world, as we have
already observed, the general tendency of its
constituent principles, seems to be toward a
state of equilibrium, or repose. But a very super-

ficial examination of those parts of the earth's crust, to which we can obtain access, is sufficient to convince us, that this quietude has not *always* existed; and consequently, that the present state of things must have had a beginning. In short, the phenomena of geology appear to show, that our earth during its progress, has undergone, alternately, periods of comparative quietude, like that in which we now live; and periods of derangement and convulsion, in which the preceding states of quietude, and their consequences, have been more or less subverted; and a new order of things, has been induced. To enter further into details regarding these changes, however, would be quite foreign to the object of the present volume. It is the business of the Geologist, to point out the changes which our earth has evidently undergone, before it arrived at its present condition; to trace the earth, as it were, from a state of chaos, through all its metamorphoses, whether sudden and convulsive, or slow and gradual; and to show, that all these changes have not resulted from chance, but from the agency of an intelligent Being, operating with some ulterior purpose; and according to certain laws, to which he had chosen to restrict himself:—to demonstrate, in fact, that to these very convulsions and changes, we owe all that boundless variety of sea and of land, of mountain and plain, of hill and valley; all that endless admixture of rocks, of strata and of soils, so essential to the existence of the present order of things; without which the world would have been a mass of crystals, or one dreary monotonous void, altogether unfit for the support of the present races of organized beings, and precluding the existence of man—apparently, one great end and object of creation. Such is the business of the geologist; and where his duties terminate, those of the Meteorologist may be said to begin. To the Meteorologist, more especially, it belongs to consider the globe in its *present* condition of equilibrium; and the means by which this state of equilibrium is maintained : in particular, to point out the influences of heat and of light; and of the energies allied to them; to study the laws of the distribution, and change of these wonderful agents, in the production of climate; to trace, in short, the effects of heat and light on the earth, the ocean, and the atmosphere; and all the infinite variety of phenomena dependent on them.

In so wide and varied a field of enquiry, it is not perhaps easy to devise a plan, that shall be perfectly unexceptionable. For, as there is no one subject so, entirely isolated, as not to be more or less influenced by the rest; we scarcely know which to commence with. After a good deal of reflection, we have adopted that arrangement, which seems to offer the most natural view of these subjects; and at the same time appears best calculated to illustrate the design, and wisdom, of the Great Creator.

CHAPTER I.

OF THE GENERAL STRUCTURE OF THE EARTH: PAR-
TICULARLY WITH REFERENCE TO THE DISTRIBU-
TION OF ITS SURFACE INTO LAND AND WATER;
AND WITH RESPECT TO ITS ATMOSPHERE.

SECTION I.

Of the General Relations of the Sea and the Land to each other.

OUR earth may be considered to be composed
of various solid, liquid, and gaseous materials;
the absolute proportions of which to each other,
we cannot even conjecture. Of the mean density of the whole, however, we can form some
estimate; and philosophers have shown, that
this density lies between five, and five and a
half, the density of water being supposed to be
one. We can also form a tolerably precise
notion, of the relative proportions of the surface,
occupied by the solid, and the liquid materials;
and of the pressure and height of the atmosphere, by which these solid and liquid materials
are surrounded.

With the general geographical distribution of land and ocean, we take it for granted, that all are more or less acquainted. We shall, therefore, confine our remarks chiefly, to their relative proportions; which are such, that nearly three-fourths of the earth's surface may be said to be covered with water; while barely one-fourth, of course, must be occupied by dry land. Of this dry land, as is well known, by far the greater part is confined to the northern hemisphere; while in the southern hemisphere, the Pacific ocean exhibits a nearly continuous surface of water, greater than that of the whole dry land of the globe put together. According to the estimate of Humboldt, the dry land in the two hemispheres, is in the ratio of three to one; between the tropics, in the two hemispheres, as five to four; and without the tropics, as thirteen to one; the preponderance being in the northern hemisphere.

The height of the dry land above the general level of the ocean is very various; but its utmost height, as compared with the diameter of the earth, is quite trifling; and it has been shown, that if the whole of the dry land existing, were equally distributed over the bottom of the sea, the quantity of water in the sea is amply sufficient to cover it entirely. Hence, " dry land can be only considered as so much of the rough surface of our globe, as may happen for the time, to be above the level of the waters; beneath which, it may again disappear; as it has done at different previous periods." *

The solid portions of our earth, are all made up of various combinations of the elementary principles, described in a former chapter. The relative situations these principles occupy in the earth's structure; the endlessly varied proportions in which they exist; and all the infinite diversity of their properties, it is the business of the geologist, and of the mineralogist to inquire into, and explain: the observations, therefore, which we have to make on the present part of our subject, will be chiefly confined to the waters of the ocean; and to the atmosphere.

SECTION II.

Of the Ocean.

THE waters of the ocean are not pure, but contain, as is well known, a variety of saline matters in solution. Indeed, when we reflect upon the immense relative extent, and general circumstances of the ocean, we may naturally suppose, that its waters will contain more or less, of every

* De la Beche's Geological Manual, p. 2.

P.   O

existing soluble principle. By far the most abundant principle, however, in sea-water, is common salt; which may be said to constitute, in general, nearly two-thirds of the whole saline matter present. The whole saline matter is between three and four per cent.; and the specific gravity of the water varies, according to the proportion of the saline ingredients, from about 1026 to 1030; pure water being supposed to be 1000. The late Dr. Marcet, some years ago, made a series of interesting experiments on this subject; and the following are the general conclusions which he drew from them :—

1. That the southern ocean contains more salt than the northern ocean, in the ratio of 1.02919 to 1·02757.

2. That the mean specific gravity of sea-water, near the equator, is 1·02777 ; or intermediate between that of the northern, and that of the southern hemispheres.

3. That there is no notable difference in sea-water under different meridians.

4. That there is no satisfactory evidence that the sea, at great depths, is more salt than at the surface.

5. That the sea, in general, contains more salt where it is deepest, and most remote from land ; and that its saltness is always diminished, in the vicinity of large masses of ice.

6. That small inland seas, though communicating with the ocean, are much less salt than the ocean.

7. That the Mediterranean contains rather larger proportions of salt, than the ocean.*

The saltness of the sea, therefore, is considerably influenced, at least at its surface, by the neighbourhood of large rivers, and by permanent accumulations of ice; and in this way, the inferior saltness of small inland seas, particularly in high latitudes, may in general be explained; as most of these inland seas are supplied with comparatively large quantities of fresh water, from the rivers flowing into them. On the other hand, the superior saltness of the Mediterranean, has been ascribed to the immense evaporation from its surface; the consequence principally, of its being situated in a warmer climate.

The saline contents of the ocean are of immense importance in the economy of nature. Such indeed is their importance, that it is doubtful whether the present order of things could be maintained without them. The effects of these saline matters, will be more particularly pointed out hereafter. In this place, we shall only remark, that by lowering the freezing point of water; and by diminishing its tendency to give

* Philos. Trans. 1819.

off vapour; they perform the most beneficial offices. Another valuable purpose which they serve, may be alluded to here; viz. the greater power of buoyancy which they communicate to water; by means of which, the waters of the ocean are better fitted for the purposes of navigation. Nor are these the only uses of the saline matters; for there is reason to believe, that they contribute in no small degree to the stability of the water; and that an ocean of fresh water would speedily undergo changes, which would probably render it incompatible with animal life; the waters of such an ocean might even be decomposed, so as seriously to interfere with the other arrangements of nature.

Lastly, who will venture to assert that the distribution of sea and of land, as they now exist, though apparently so disproportionate, is not actually necessary, as the world is at present constituted? What would be the result, for instance, if the Pacific or the Atlantic oceans were to be converted into continents? Would not the climates of the existing continents, as formerly observed, be completely changed by such an addition to the land; and the whole of their fertile regions be reduced to arid deserts? Now, this distribution of sea and of land, so wonderfully adapted as it appears to be to the present state of things, depends of course in a great measure, upon the *absolute quantity* of water in the world. While on the other hand, the *relative gravity* of water, as compared with that of the earth, keeps the ocean within its destined limits, notwithstanding its incessant motion. Thus Laplace has shown, that the world would have been constantly liable to have been deluged from the slightest causes, had the mean density of the ocean *exceeded* that of the earth ! Hence the adjustment of the quantity of water, and of its density, as compared with that of the earth, afford some of the most marked, and beautiful instances of design.

SECTION III.

Of the Atmosphere.

THE immense body of gaseous matters surrounding our earth, and usually known under the name of the Atmosphere, is essentially composed, as we formerly stated, of two principles, oxygen and azote, in the proportion nearly of one part of oxygen, and four parts of azote. Besides these two gases, the atmosphere also contains a small, and perhaps a variable, quantity of carbonic acid gas, amounting upon an average, to somewhat less than one part in a thousand of the whole ; and of water in a state of vapour, likewise a variable quantity, (as will be shown hereafter), but usually fluctuating between one, and one and a half per cent.*   In addition to these ingredients, there are, probably, also other matters constantly present in the atmosphere; for as the sea contains a little of every thing that is soluble in water; so the atmosphere may be conceived to contain a little of every thing that is capable of assuming the gaseous form.

The atmosphere exerts a pressure, or weight, upon all parts of the earth's surface, on an average, equal to about fifteen pounds upon a square inch; or in other words, equal in weight to a column of mercury, one inch square, and thirty inches high. The well-known instrument, the common *Barometer*, or *Weather-glass*, consists of nothing more, than such a column of mercury, poised or pressed upwards into a vacuum, by the weight of the atmosphere. With the changes constantly taking place in the height of such a column, every body is familiar; and we shall have occasion to recur to them hereafter: at present, it is only requisite to observe, that these changes are much less remarkable in tropical,

---

* Or, more accurately speaking, 1000 parts of atmospheric air, under ordinary circumstances, may be said to consist of

Oxygen	210·0
Azote	775·0
Aqueous vapour . . .	14·2
Carbonic acid	0·8
	1000·0

than in temperate climates. Thus, between the tropics, the barometer usually varies only about one-third of an inch ; while in temperate cli-mates, the changes amount to upwards of one-tenth of the whole height,

The pressure of the atmosphere decreases as we ascend above the earth's. surface ; and for equal ascents, this decrease of density, is, in what is called, geometrical progression, Thus, after an ascent of three miles, the density of the atmosphere is found to be only one half of what it is at the surface of the earth, or equal to a column of mercury fifteen inches in height ; at six miles, the barometer would stand at one-fourth of its usual height, or seven and a half inches ; at nine miles of elevation, at three inches and three quarters ; and, at fifteen miles, nearly at one inch only. Hence, though from various circumstances, the atmosphere has been inferred to extend from forty to forty-five miles above the earth's surface ; by far the greater portion of it is always within fifteen or twenty miles. The distance, however, to which the atmosphere extends, must be different in different latitudes ; for the rotation of the earth upon its axis ; and the greater, and more direct influence of the solar heat near the equator, will necessarily cause the atmosphere to be higher in the equatorial, than in the polar regions ; while at the poles, the atmosphere must be lower, than over any other part of the earth's surface.

Much difference of opinion has existed among philosophers, as to the mode in which the various principles, entering into the composition of atmospheric air, are associated ; some maintaining that these principles exist simply in a state of mixture : others considering them as chemically united. We formerly stated that all gaseous bodies, when they combine with one another, combine with reference to their volumes ; that is to say, that one volume of one gas always combines with one, two, or more similar volumes of the same, or of another gas, and not with any intermediate fractional part. Now, since atmospheric air is essentially composed of one volume of oxygen, and four volumes of azote, it is evident, whether its elements be in actual union or not, *that it is at least constituted upon strictly chemical principles ;* whence it follows, that the composition of the atmosphere has not been the result of accident. In this point of view, therefore atmospheric air may be considered to be as much a chemical compound as water, or any other similar body ; and instead of viewing the atmosphere, according to a prevalent notion, as a mere accidental and heterogeneous appendage, connected with the denser matters by no apparent tie ; we may fairly rank the atmosphere among the constituent principles of our globe ; and as forming a symmetrical part, of the great harmonious whole.

But although atmospheric air has been thus originally constituted upon chemical principles, and probably owes its stability, in no small degree, to this circumstance; yet the mode in which its constituent elements are associated, is very different from that, in which the elements of compounds in general, are associated. Indeed the constituent elements of atmospheric air, do not appear to be combined at all; but to be only mixed, or simply diffused through each other, in the same manner, as the minute portions of carbonic acid gas, and of vapour, are known to be diffused through the whole atmosphere; that is to say, according to the laws of the general diffusion of gaseous bodies, which we endeavoured to explain in a former chapter. To this explanation we must refer the reader for details. We shall merely observe here, that the fundamental principle of this explanation consists in the assumption, that the molecules of all bodies in the gaseous state, are self-repulsive, (or repulsive of one another, in preference to others), for the same reason, that in the solid state, they are self-attractive, (or attract one another, in preference to others). When different gaseous bodies therefore, are mixed together, they will not assume a position according to their specific gravities, as they might otherwise be expected to do; but the molecules of each gas, will be equally diffused throughout the whole space

occupied by the mixture. Hence, one direct
and most important effect of the mixed constitution of the atmosphere, is *its nearly uniform
composition*, at least within the limits attainable
by man :—a fact which has been confirmed by
innumerable analyses of the air, made in all parts
of the world; both at its surface, and at the greatest
heights man has hitherto reached. Moreover,
this constitution of the atmosphere, not only originally produced such uniformity of composition;
but *is the cause constantly operating to preserve
that uniformity*—the grand conservative principle,
as it were, preventing any unequal distribution
of the constituent elements of the atmosphere;
which would speedily prove fatal to organic life!
Were the gaseous principles composing the atmosphere in ever so slight a state of union, they
could not readily diffuse themselves through
each other; and partial accumulations of one or
other of them would be constantly taking place:
but as the atmosphere is at present constituted,
if a little more oxygen be consumed in one spot
than in another; instantly the deficiency is supplied from the neighbourhood by diffusion; and
the equilibrium is scarcely affected, in a sensible
degree. Another curious result of this independent condition of the gaseous principles of the
atmosphere is, that of the whole pressure exerted, each principle exerts its own force, according to its quantity. Thus, of the thirty inches of

mercury supported by the whole atmospheric pressure, the azote sustains 23 $\frac{36}{100}$ inches, and the oxygen 6 $\frac{13}{100}$ inches; while the aqueous vapour sustains only $\frac{44}{100}$ inch, and the carbonic acid still less, or only $\frac{3}{100}$ inch. Hence it is evident, that the fluctuations in the height of the barometer, (amounting to nearly three inches in our latitude,) cannot depend altogether upon the quantity of aqueous vapour in the atmosphere; for if the whole of this vapour were annihilated, it would scarcely produce a difference in height of half an inch. Attention is now drawn to this fact, for purposes, which will appear in a subsequent chapter.

Lastly, had the absolute quantity, or the relative gravity, of the atmosphere, been materially different from what they are; the present order of things could not have existed. Hence, the same striking evidences of wise adjustments are displayed, in these arrangements of the atmosphere, as in the arrangements formerly shown to exist, with respect to the quantity and the gravity of the waters of the ocean.

Before we close the present chapter, let us reflect for a moment, upon the great arrangements we have been considering.

Why has the surface of this earth been divided into land and sea? Why have the land and sea been so adjusted to each other, that their condition and proportions hardly admit of change,

without destruction to the whole fabric? Why
has their present stability been so wonderfully
secured? Again, with respect to the atmosphere;
why has any atmosphere been thrown around
this globe? and why such manifest provisions to
secure its ubiquity, and unvarying constitution?

Viewed alone, and without reference to organized beings, all these things appear to want an
object. This globe might have revolved about
the central luminary—might have occupied its
point in the universe, without any " gathering
together of the waters,"—without any circum-
ambient air. But the scheme of the great Creator extended beyond the mere adaptation of
inanimate matter. " Before its foundations were
laid," He had destined this earth to teem with
life; and throughout, has displayed his original
design of rendering it a fit habitation for living
beings. For this purpose, and acting, at the
same time, in strict conformity to those laws, by
which He had chosen to limit himself, He has,
by means of successive convulsions and changes,
so contrived to mix and blend the different elements; and finally, so to arrange the dry land
apart from the sea; that, taken as a whole, and
with reference to the present order of things,
their relative proportions will scarcely admit of
material change. While, to crown his works,
and as it were, the more strongly to evince his
design and his wisdom, He has surrounded this

globe with an atmosphere ; to preserve the homogeneity of which, its principles have been so associated, as to constitute an exception to his usual operations, and even to the general laws of nature!

CHAPTER II.

OF HEAT AND LIGHT——THE MODES OF ESTIMATING THEIR DEGREE ; AND THE WAYS IN WHICH THEY ARE PROPAGATED. OF THE GENERAL TEMPERATURE OF THE CELESTIAL REGIONS ; AND OF THE EARTH ; INDEPENDENTLY OF THE SUN.

SECTION I.

Of Heat and Light; and of the Modes of estimating their Degree.

OUR sensations are a very imperfect measure of temperature ; and when we wish to speak with precision on that subject, it becomes necessary to have recourse to other means of comparison. For the sake of the general reader, we shall, therefore, in the first place, briefly describe the principles of the construction of the *Thermometer*, the instrument for measuring heat.

All bodies, as we have shown in a former

206   METEOROLOGY.

chapter, become more or less expanded, when they undergo an increase of temperature. Hence, the relative degrees of expansion of any body, may be viewed as a sort of measure of the degree of heat; and most of the thermometers employed, act upon this principle. Thus the common thermometer, as is well known, consists of a portion of some fluid, generally of mercury, enclosed in a small glass ball; the cavity of which ball communicates with a tube of narrow bore. We shall suppose the quantity of the mercury, and the size of the ball, to be so adjusted to each other; that when the instrument is placed in ice on the one hand, and in boiling water on the other; the whole expansion of the mercury between these two fixed temperatures, shall fall within the range of the tube. The points at which the mercury stands in the tube, at the freezing, and boiling temperatures, are to be accurately noted; and the intermediate space upon the scale attached to the tube, is to be divided into 180 equal parts or degrees; the freezing point is to be marked 32°, and of course, the boiling point 180° above, or 212°. Such is *Fahrenheit's* scale, the one employed in this country, and to which, the numbers hereafter mentioned refer. In other countries different scales are made use of; thus in Sweden, France, and elsewhere, what is termed the *centigrade* thermometer is generally adopted. In this thermometer, the freezing point is marked 0°, and the boiling point 100°. In other parts of the continent, *Reaumur's* scale is much used. In Reaumur's scale, the freezing point, as in the Centigrade, is 0°; but the boiling point is only 80°. These different graduations are easily convertible; but it is much to be regretted that they exist, as they cause considerable trouble and confusion.

The instrument employed for measuring the intensity of light, is termed a *Photometer;* of such an instrument various forms have been proposed, but at present they are all very imperfect.

SECTION II.

Of the Propagation of Heat and Light.

THE modes in which heat and light are propagated from one body to another, and through the same body, have been already explained, and we need not again enter into details: a brief recital here, however, of the modes in which heat and light are actually propagated among the objects of nature, may not be unacceptable to the general reader.

Heat passes from the sun to the earth by

radiation; and again, by the same process, it is freely sent off from the surface of the earth into the atmosphere. Below the surface of the earth, heat is propagated in all directions through the *solid* matter, by what is called *conduction.* A third mode in which this important agent is extensively propagated in nature, is by the means we have termed *convection,* or the *carrying* process. Convection is confined, of course, to fluids, as water and air. A portion of water or of air being heated above, or cooled below the surrounding portions, expands or contracts in magnitude, and thus becoming specifically lighter or heavier, rises or sinks accordingly; carrying with it, the newly acquired temperature, whatever that temperature may be.

Light, at present, is only known to be propagated by radiation.

By bearing in mind these modes of the propagation of heat and light; the general reader will find no difficulty in understanding what follows.

SECTION III.

Of the Temperature of the Celestial Regions.

·FROM the close and intimate relations between heat and light, and from their almost invariable association as they exist around us, it seems not very unreasonable to conclude, that these agencies are generally associated in nature; and that wherever one is present, there the other must be present also. If this be really the case, the innumerable fixed stars, considered to be so many suns, must be supposed capable of diffusing heat, as well as light, throughout the celestial regions; and consequently there must be a certain degree of temperature, common to the whole. For this reason, and for others which might be mentioned, philosophers have not only inferred the existence of such a common temperature throughout the celestial regions, independently of our sun; but have even attempted to determine its degree. Moreover, all the different modes which have been employed to estimate this temperature, singularly coincide in showing, that it does not differ much from — 58° of Fahrenheit's scale. The temperature of space is, therefore, supposed to be about 90° below the freezing point of water;

P.   P

a degree of cold " not greatly inferior to that at which quicksilver becomes solid ; and much superior to some degrees of cold which have been produced artificially." * If such a common temperature do indeed exist throughout space, or at least in our planetary system, it must have no inconsiderable influence upon the temperature of the planets generally ; and with respect to our own globe in particular, such a common temperature must operate, by diminishing the intensity of the cold around the poles.

SECTION IV.

Of the Temperature of the Interior of the Earth.

THE attention of philosophers has, for some years past, been a good deal directed to the internal temperature of the earth, at great depths ; beyond the influence of the sun, or of any other external cause. From the earliest times, some vague notions of a central heat seem to have existed among mankind ; doubtless, arising from their attention being forcibly drawn to the phenomena of volcanoes, and hot springs ; but it is not till a comparatively late period, that

* Discourse on the Study of Natural Philosophy; p. 157. By Sir J. F. W. Herschel.

the subject has been carefully investigated. It would be quite foreign to our present purpose to enter here into details; we shall therefore merely state, that the arguments in favour of the probability of a central heat, are—" first, the experiments made in mines, which, notwithstanding their liability to error from various sources, still seem to show, particularly those made in the rock itself, an increase of temperature from the surface downwards ;—secondly, the existence of thermal springs, which are not only abundant among active and extinct volcanoes, but also among all varieties of rocks in various parts of the world ;—thirdly, the existence of volcanoes themselves, which are distributed over the globe, and present such a general resemblance to each other, that they may be considered as produced by a common cause, and that cause, probably, deep-seated ;— and lastly, the terrestrial temperature at comparatively small depths, which does not coincide with the mean temperature of the air above it."*

Such is an abstract of the principal arguments which have been brought forward in support of the opinion, that within our earth, even at the present time, there exists a central heat of great intensity. As corroborative of the same

* De la Beche's Geological Manual, p. 24, new edit.

views, may be mentioned the evidence derived
from the characters of the fossil remains both
of plants and of animals, found in the colder
regions of the world ; which characters are such,
as to prove beyond a doubt, that these plants
and animals must have existed in a climate
much hotter than that in which their remains
are found ; and indeed, of equal, if not of superior heat, to that of the tropical portions of our
earth at the present time. Hence it has been inferred, that the temperature of our earth, formerly
much above what it is now, has been gradually
dissipated into the surrounding planetary regions, and thus helped to increase the general
temperature, above stated, as supposed to exist
throughout space. Moreover, the Baron Fourier,
to whom we are principally indebted for these
observations, has attempted to show, that the
earth has nearly reached its limit of cooling,
particularly near the surface. Near the surface,
the temperature would necessarily decrease
much more rapidly than in the interior ; where,
in a globe of the earth's magnitude, the temperature might be supposed to remain nearly
unchanged, for a very great length of time.
The same distinguished philosopher has also
attempted to show, that the temperature of the
surface is still liable to be influenced, by the
gradual escape of heat from the interior, which

even yet seems to be constantly going on; and
that the temperature of the surface is thus somewhat higher, than it would be, if such a central
heat did not exist; or than if the temperature
of the surface of the earth depended only;
upon the action of the sun. We are thus brought
to the proper commencement of this treatise
on Meteorology; viz. the consideration of the
present state of the earth's temperature, as liable
to be influenced by the presence or absence
of the sun, the great source of heat and of life
to our system.

Before proceeding, we may remark, that the
details of the subject we have now concluded,
fall entirely within the province of the geologist.
To him it belongs, as we have already said,
not only to trace the wonderful changes which
our globe has undergone in arriving at its
present condition; but to point out, the beautiful adaptations of organic life, and structure,
to the existing circumstances of its various
epochs. Considered in this point of view,
geology is a subject of the highest interest and
importance; and, to use the words of an emi-
nent Professor, with which we shall finish this
chapter, " lends a great and unexpected aid
to the doctrine of final causes; for it has not
merely added to the cumulative argument, by
the supply of new and striking instances of

mechanical structure adjusted to a purpose, and that purpose accomplished; but it has also proved, that the same pervading active principle manifesting its power in our times, has also manifested its power in times long anterior to the records of our existence.

" But, after all," continues our author, "some men, seeing nothing but uniformity and continuity in the works of nature, have still contended (with, what I think, a mistaken zeal for the honour of sacred truth) that the argument from final causes proves nothing more than a quiescent intelligence. I feel not the force of this objection. In geology, however, we can meet it by another direct argument; for we not only find in our formations organs mechanically constructed, but at different epochs in the history of the earth, we have great changes of external conditions, and corresponding changes of organic structure; and all this, without the shadow of a proof, that one system of things graduates into, or is the necessary and efficient cause, of the other. Yet in all these instances of change, the organs, as far as we can comprehend their use, are exactly those which were best suited to the functions of the being. Hence we not only show intelligence contriving means adapted to an end, but, at successive times and periods, contriving a change of mechanism adapted to a change in external conditions. If this be not the operation of a prospective and active intelligence, where are we to look for it ? " *

CHAPTER III.

OF THE TEMPERATURE OF THE EARTH AT ITS SUR-FACE, AS DEPENDENT ON THE SUN.

THE general temperature of the earth is doubtless regulated by its situation in the universe ; and more especially, by its position with respect to the sun. To this position, as formerly observed, the properties of its constituent principles have, most obviously, been all adapted with consummate wisdom ; so that, under the circumstances in which they are placed, some are solid, some liquid others gaseous, according to the purposes they are intended to fulfil in nature.

But the heat and light derived from the sun, are very unequally distributed over the surface of the earth ; and every one is familiar with the fact, that as we recede from the equator towards

* Address delivered to the Geological Society of London, by the late President, Professor Sedgwick, 1831. -

the north or south, the temperature of the earth's surface gradually diminishes, till we arrive at the polar regions.

Such is the general fact. But the circumstances which conspire to interfere with this gradual distribution of temperature, are so numerous and so influential ; that the actual temperature of a place can be learnt only by observation. Among the circumstances thus more especially affecting the distribution of temperature, may be mentioned, the nature of the surface, whether water or land ;—and the situation, whether at a greater, or at a less height, above the level of the ocean. To such circumstances may be added, the particular configuration and geographical relations of places : as their aspect to the north or south ; their being sheltered or exposed ; the composition and nature of the soil, particularly its colour and state of aggregation ; on which depend its powers of absorbing and of radiating heat and light ; and of retaining or of parting with humidity, &c. ; also the proximity, or absence of, seas ; the predominancy of certain winds ; the frequency of clouds, fogs, &c. These, and innumerable other circumstances, many of which will be pointed out in subsequent chapters, contribute to influence the temperatures of different places ; and to render them, in fact, as varied as the places themselves.

Nor is difference of place, the only cause of difference of temperature; every one knows, that at the *same* place, the temperature is in a constant state of change. Hence, before we can obtain correct notions of the actual temperature of any given place, or period, certain expedients are necessary, which it will be requisite first to consider.

SECTION I.

Of Mean Temperature.

IF, on any given day, we observe the temperature at the earth's surface, at the commencement of every one of the twenty-four hours, we shall find, as before observed, that at each hour the temperature is different; and we naturally enquire, which of all these temperatures is to be chosen in preference, as the one characteristic of the day and place? The answer to this question obviously is; *that temperature, whatever it may be, which is equidistant from the extremes;* or, as it is usually termed, the *mean temperature of the whole.* Now this mean temperature may be obtained, nearly, by adding all the results together, and dividing the sum by the number of obser-

vations; thus we arrive at the mean temperature
of the day, by adding together the temperatures
observed at different hours of the day, and
dividing the sum by the number of temperatures. In like manner, by adding together the
mean temperatures of every day of a week, or
of a month, and dividing the sum by the number
of days, we obtain the mean temperature of the
week or month; and so on, by similarly treating
the mean temperatures of the months, or of any
number of years, we obtain the *mean* temperature of the year, at a given place: and it is to
be remembered, that the more numerous the observations, the more accurate will be the mean
result.

Lastly, it remains to state, that the temperature
always understood by the Meteorologist, (unless
otherwise expressed), is the temperature of the air
near the surface of the earth, as indicated by a
thermometer, effectually protected from radiation
and foreign influence of every kind. The temperature as indicated by a thermometer fully
exposed to solar radiation, and which in its turn
is allowed to radiate freely in the sun's absence,
is altogether a different thing; and may be imagined to coincide very nearly with the actual
temperature of the earth's surface, when similarly exposed. The fluctuations of temperature
indicated under these circumstances, are much
greater than those of the air above noticed;

though it is probable, that the mean of the whole of such observations, if this mean could be accurately obtained, would differ little from the mean of those of the air.

SECTION II.

Of the actual Distribution of Temperature over the Earth. Of Isothermal Lines, &c. CLIMATE.

THE reader is supposed to be acquainted with the principles of the common division of the surface of the globe into five zones or portions, usually denominated the *torrid*, the two *frigid*, and the two intermediate, or *temperate zones;* and that generally speaking, the poles, and the equator, present the extremes of temperature upon the earth's surface. Now, in considering the general distribution of temperature over the globe, the extreme temperatures naturally claim our attention in an especial manner : we shall, therefore, in the first place, proceed to consider the temperature of the polar, and of the equatorial regions.

Of the Temperature of the Poles, and of the Polar Regions.—The probable mean temperature of the poles has always been an interesting subject of meteorological enquiry. It must be confessed, however, that after all that of late years has been done by our enterprizing countrymen, much is yet necessary, to enable us to arrive at satisfactory conclusions. Thus it has been shown, that in attempting to calculate the temperature of the North Pole, we shall obtain very different results, by employing the temperature occurring in the old world, and that observed in the new world; the temperature of the old world indicating the temperature of the pole to be about 10°; while the temperature of the new world, indicates it to be considerably below Zero. Hence it has been inferred, that there are two points or poles of greatest cold, situated in about the latitude of 80° north, and in longitudes 95° east, and 100° west; and consequently, that the geographical pole of the globe, is not the coldest point of the Arctic hemisphere. Whether this deduction be well founded or not, must be decided by future observation. At present, the actual temperature of the Polar regions cannot be considered as determined.

Although we are thus unable to state with certainty the temperature of the Polar regions, it may nevertheless be deemed an object of curiosity, to know the *lowest* temperatures that have been noticed. Perhaps the lowest *authentic* observations of temperature we possess, are those by Captain Parry at Melville Island.

There, the thermometer in the ship, was often observed as low as—50°; and at a distance from the ship, even as low as 55° under Zero. We believe still lower temperatures than these are on record, but probably they are not to be relied on. The greatest degree of cold hitherto produced *artificially*, has been 91° under Zero.

Of the mean annual Temperature of the Equator. The mean annual temperature of the equatorial, like that of the polar regions, is a meteorological problem of considerable interest. Humboldt, from a very extensive generalization, fixed the mean equatorial temperature at 81½°; and the same temperature has been adopted by others. Attempts, however, have been recently made to show that this temperature is 3° or 4° below the truth; but Humboldt in reply still maintains his former opinion. Since at the equator, only about one-sixth of the whole circumference of the globe is dry land; the general equatorial temperature, as actually found to exist, is perhaps lower than upon theoretical principles it ought to be; and certainly much below what it ought to be, as deduced from observations made on the continent in the neighbourhood of the equator. Thus the mean temperature of Pondicherry, in latitude 11° 55' north, is at least 85°; and if from this temperature, the temperature of the equator were deduced according to the common principles, the deduction would of course be much above the truth. The fact is, as in the case of the Polar regions, we do not possess the requisite data for determining the equatorial temperature, in a perfectly satisfactory manner.

As in speaking of the Polar regions, we noticed the *lowest* degree of temperature which had been observed; perhaps, while speaking of the equatorial regions, it may be deemed not irrelevant, to notice the *highest* temperature. Observations, however, of this kind, being principally founded on the incidental notices of travellers, are not, in general, much to be relied on ; or are to be considered only, as approximations. Thus the thermometer has been recorded at Benares to stand at 110°, 113°, and even 118°. At Sierra Leone, it has been observed, when placed on the ground, to indicate a temperature of 138°. Humboldt also gives many instances of the temperature of the surface of the earth, amounting to 118°, 120°, and 129° : and on one occasion he found the temperature of a loose and coarse granitic sand, to amount to upwards of 140° ; the thermometer in the sun at the time, only indicating a temperature of about 97°.

Of the Temperature of the intermediate Regions of the Globe. Of Isothermal Lines, &c. With respect to the temperatures of those parts of the earth, between the poles and the equator, it may be remarked, that, except for reference only, the old division, before mentioned, of the earth's surface into zones, is now almost entirely superseded by the more precise and natural arrangement, termed the *Isothermal* arrangement. According to this arrangement, all the places upon the globe, having *the same annual mean temperature*, are classed together ; and lines drawn upon a map through such a series of places, have been termed *Isothermal lines*, or *lines of equal temperature*. As might be expected from what has been already stated, the courses of these lines are by no means regular. Thus, suppose two travellers set out, the one from London and the other from Paris ; and each visit all the places in the northern hemisphere, in which the mean annual temperatures are the same as in these two cities. It will be found that the lines of their routes, or the isothermal lines of these two cities, will not only not follow the parallels of their latitude, but that they will not be parallel to each other ; and the same may be said to be the case, with any other two places upon the globe. Hence, as the isothermal lines are as numerous as the places, and as diversified as numerous, geographers have grouped them into bands or zones. Thus Humboldt, to whom we owe most of what has been done on this subject, has divided the northern hemisphere into the following six isothermal bands, or zones, viz.

1. The zone of mean annual temperature ranging from 32° to 41°.
2.   -   -   ..   -   from 41° to 50°.
3.   -   -   -   -   from 50° to 59°.
4.   -   -   -   -   from 59° to 68°.
5.   -   -   -   -   from 68° to 77°.
6.   -   -   -   -   from 77° upwards.

The tables given in the appendix contain a general view of Humboldt's results. From these, and from other data the approximate courses of the different isothermal lines have been traced on the accompanying map; which will convey to the reader a much more distinct notion of their nature, than can be conveyed by words. We shall therefore content ourselves with briefly pointing out the approximate course of the most interesting of these lines; viz. *the Isothermal line of* 32°.

If we begin to trace this important line from the eastern parts of Siberia in longitude 130° east, we shall find that in that meridian, it commences nearly in the latitude of 59° north ; whence it makes a gradual bend northwards, and crosses the parallel of 60°, nearly in longitude 90°. From that point, it still advances to the northward, and crossing the arctic circle in longitude 45° east, arrives at its most northern extremity in about latitude $67\frac{1}{2}$°, longitude 10° east. From this, its most northerly limit, the line takes a gradual sweep towards the south ; recrosses the arctic circle in longitude 15° west, and passing through the north-west of Iceland, divides the parallel of 60°, in longitude 42° west. Thence the line proceeds southwards to the latitude of 54°, a little to the north of Table Bay, in Labrador; gradually declining in its course till it arrives at longitude 100° west, in the central parts of the new continent. The Isothermal line of 32°, ranges, therefore, through a space of 14° or 15° of latitude; while its western extremity, in the central parts of America, is 5° or 6° nearer the equator, than its eastern extremity in Siberia—a circumstance strikingly illustrative of the greater cold of the new continent, in the same parallel of latitude. The other Isothermal lines are represented approximately on the map, and do not require to be more minutely described. The most remarkable circumstance connected with them is, that, as they approach the equator, they gradually become less convex towards the north ; so that the Isothermal line of 77° differs but little from a straight line, coincident with the tropic of cancer.

In the arrangement above described, the mean temperatures of *the whole year* are supposed to be classed together ; but it is obvious that the same principle may be applied to any portion of the year ; as the extreme winter, and summer, temperatures. Such classifications are often,

P.   Q

as we shall presently see, of great importance, in enabling us to estimate the characters of a particular country. Lines drawn through places having the same summer, and the same winter, temperatures, are denominated *Isotheral* and *Isocheimal* lines; while lines drawn through places having other common temperatures, receive other appropriate names.

After these general remarks, we proceed to give a summary sketch of the actual distribution of temperature over the *northern hemisphere*, which we shall subjoin in the words of Humboldt.

" The whole of Europe," says this distinguished philosopher, " compared with the eastern parts of America and Asia, has an insular climate; and upon the same Isothermal line, the summers become warmer, and the winters colder, as we advance from the meridian of Mont Blanc towards the east or the west. Europe may be considered as the western prolongation of the old continent; and the western parts of all continents are, not only warmer, at equal latitudes, than the eastern parts; but even in the zones of equal annual temperature, the winters are more rigorous, and the summers hotter, on the eastern coasts, than on the western coasts, of the two continents. The northern part of China, like the Atlantic region of the United States, exhibits seasons strongly contrasted; while the coasts of

New California, and the embouchure of the Columbia, have winters and summers almost equally temperate. The meteorological constitution of countries towards the north-west, resembles that of Europe as far as 50° or 52° of latitude. Comparing, in the two systems of climates, the concave and the convex summits of the same Isothermal lines; we find at New York, the summer of Rome, and the winter of Copenhagen; at Quebec, the summer of Paris, and the winter of St. Petersburgh. At Pekin, also, where the mean temperature of the year is that of the coasts of Brittany, the scorching heats of summer are greater than at Cairo, and the winters are as rigorous as at Upsal. So also, the same summer temperature prevails at Moscow. in the centre of Russia, as towards the mouths of the Loire, notwithstanding a difference of 11° of latitude; a fact that strikingly illustrates the effects of the earth's radiation, on a vast continent deprived of mountains. This analogy between the eastern coasts of Asia and America sufficiently proves," continues Humboldt, " that the inequalities of the seasons, depend on the prolongation and enlargement of continents towards the pole; on the size of seas in relation to their coasts; and on the frequency of the north-west winds; and not on the proximity of some plateau, or elevation, of the adjacent lands. The great table lands of Asia do not stretch

beyond 52° of latitude ; and in the interior of the
new continent, all the immense basin, bounded
by the Alleghany range, and the rocky mountains, is not more than from 656 to 920 feet
above the level of the ocean."

The following remarks apply to *the tempera-*
ture of the southern hemisphere.

The general temperatures of the northern, and
of the southern hemispheres, are understood to
differ very considerably. This difference, however, does not depend upon any material difference in the proportion of heat and light derived
from the sun, as will be presently shown ; but
on the very unequal distribution of sea and of
land, in the two hemispheres ; the small quantity of land in the southern hemisphere, contributing to equalize the seasons.

Humboldt has shown, that near the equator,
and indeed so far south as 40° or 50°, the
similar Isothermal lines are in both hemispheres almost equally distant from the poles ;
and that, in considering only the transatlantic
climates between 70° and 80° of west longitude,
the mean temperatures of the year, under corresponding geographic parallels, are even greater
in the southern than in the northern hemisphere.
It is the *division* of heat, therefore, between the
different seasons of the year, rather than the
absolute amount of heat during the whole year,
which gives a particular character to southern

climates; and approximates them generally, to the character of insular climates. The mean temperature is not precisely known beyond 51° of south latitude; yet there is no reason to believe, that the Isothermal line of 32° is much further from the south pole, than, in the opposite hemisphere, the similar line is from the north pole: and some circumstances at first sight appear to show, that the Isothermal line of 32° is even nearer to the south pole, than it is to the north pole; though these circumstances are probably deceptive. With respect to the temperature of the south pole itself, like that of the north pole, we have no means of forming an accurate estimate.

Such is a summary account of the general distribution of temperature, over the northern and southern hemispheres. Now, amidst the infinite changes every where going on, there is nevertheless, at the same place, a certain average state of things, which, taken together, constitute what is called the CLIMATE of the place. Of climate, undoubtedly, temperature is the most important ingredient. But the circumstances, besides mere temperature, which enter into the formation of climate, are so numerous and diversified; and their operation, in consequence, is so complicated; that it becomes exceedingly difficult to unravel, and display them in a satisfac-

tory manner.   The constituents of climate, however, appear to be most naturally divided into
two great sections; viz. *those of a* PRIMARY *kind,
depending upon the globular figure of the earth;
upon its motion in its orbit, and upon its axis: and
those of a* SECONDARY, *or subsidiary kind, more
immediately connected with the globe itself, and
depending upon the nature of its surface, as composed of land or water;* or, *as connected with its
atmosphere.*   Under these two divisions, we purpose to consider the subject of CLIMATE, in the
following chapters.

CHAPTER IV.

OF THE PRIMARY CONSTITUENTS OF CLIMATE: OR,
OF THE TEMPERATURE OF THE EARTH, AS DE-
PENDENT ON ITS GLOBULAR FORM ; AND ON ITS
ANNUAL AND DIURNAL MOTIONS.

THE distance of the earth from the sun is such,
that the solar rays may be supposed to arrive
at the earth's surface in a state of parallelism.
Now, when parallel rays fall upon a globe, it is
obvious, that any number of such rays falling
perpendicularly, as at the equator of our earth,
will occupy a very different portion of the surface of the globe ; from what an equal number

of the same rays will occupy, where they fall obliquely, as in our polar regions. Hence, as we recede from the equator towards each pole, heat and light are diffused. over gradually increasing portions of the earth's surface; and thus the intensity of both decreases in a like proportion. The exact law of such decrease is well known to mathematicians, but need not be here repeated. For our present purpose it is sufficient to observe, that among the natural causes affecting the distribution of heat and light in different latitudes, the globular figure of the earth is the principal.

The second great natural cause of the unequal distribution of heat and light over the earth, is the obliquity of the earth's motion in its orbit, with respect to the plane of its equator. From this obliquity it happens, that, during the annual revolution of the earth round the sun, every part of its surface, between the latitudes of $23\frac{1}{2}°$ north and south from the equator, is in turn exposed to the perpendicular influence of the sun. To this oblique motion of the earth in its orbit, we owe the endless variations and vicissitudes of seasons in different latitudes.

There is also another circumstance connected with the earth's motion in its orbit, which, as partaking of the character of a primary cause, may here be briefly noticed. The earth's orbit is not a circle, but an ellipse, of which the sun occupies one of the *foci*. Now, it has been so arranged, that in the middle of our winter, the earth is in that part of its orbit, which is nearest to the sun. The earth, therefore, is, at Christmas, actually about three millions of miles nearer to the sun, than at Midsummer. Hence it might be inferred, that the temperature of the southern hemisphere, which during our winter is directly exposed to the sun, would be affected by this greater proximity. Such, however, is not the case; for this greater proximity to the sun, is almost exactly counterbalanced, by the swifter motion of the earth along this part of its orbit. The eccentricity of the earth's orbit, therefore, has little or no influence on its temperature, as at first sight might be supposed.*

The third great natural cause affecting the distribution of heat and light over the earth,

---

* Or, to quote the more precise explanation of Sir J. Herschel, " The momentary supply of heat received by the earth from the sun, varies in the exact proportion of the angular velocity, that is of the momentary increase of longitude. Hence the greater proximity of the sun in the winter, is exactly compensated for, by the earth's more rapid motion; and thus an equilibrium of heat is, as it were, maintained. Were it not for this, the eccentricity of the orbit would materially influence the transition of the seasons; and the effect would be, to exaggerate the difference of summer and winter in the southern hemisphere, and to moderate it in the northern; thus producing a more violent alternation of climate in the one hemisphere, and an approach to perpetual spring in the other. As it is, however, no such inequality subsists; but an equal and impartial distribution of heat and light is accorded to both." Treatise on Astronomy, p. 198, (Lardner's Cyclopædia).

is the earth's revolution on its axis. To this revolving motion, we owe the innumerable minor vicissitudes of temperature, and of light and shade, daily and hourly experienced throughout the world.

Such are the three great natural causes which regulate the distribution of heat and light over our globe. They may be considered as the necessary results of more general laws, to which the Great Author of nature has chosen to restrict himself; and to which, as usual, He most rigidly adheres. Why, among the numerous possible means by which heat and light might have been, and in other instances, are, distributed from a central sun over a distant planet; these regulating causes have been selected for our earth, is absolutely unknown to us. That this selection has been made with some ulterior view, we cannot hesitate to believe; and one such view or purpose, may have been, to demonstrate to us His wisdom and His power; by the methods chosen, for obviating the difficulties necessarily resulting from these primary arrangements. In other planets, where other primary arrangements for the distribution of heat and light have been adopted; there are probably other modes of obviating the difficulties arising from them. Of such arrangements we can form no conception; but to the inhabitants of these planets, they are doubtless an equal evidence of the wisdom and the power of the Deity.

CHAPTER V.

OF THE SECONDARY, OR SUBSIDIARY CONSTITUENTS
OF CLIMATE : COMPREHENDING A SKETCH OF
THOSE CIRCUMSTANCES CAPABLE OF INFLUENCING
CLIMATE, WHICH ARE MORE IMMEDIATELY CON-
NECTED WITH THE SURFACE OF THE EARTH, AS
CONSISTING OF LAND OR WATER ; OR WHICH ARE
CONNECTED WITH THE ATMOSPHERE.

IN the preceding chapter we have alluded to the difficulties, or exigences necessarily arising from the modes in which heat and light are distributed over our earth ; and of these, before we proceed, it may be proper to specify some of the most striking.

Had the heat and light derived from the sun to the earth, not been in any way modified; the equatorial and the polar regions would have been alike inaccessible to organic life. The heat within the tropics, and the cold towards the poles, would both have been destructive ; while the intermediate regions would have been exposed to a constant succession of violent and sudden alternations of temperature, which would have rendered the present state of things no less an impossibility. In order, therefore, to render this earth an appropriate dwelling-place for such beings as at present occupy its surface, it was necessary that these extremes, and sudden vicissitudes of temperature should be in some way diminished or alleviated. Accordingly, these objects have been effected with the most consummate wisdom. Indeed, some of the most splendid instances of design in nature, are offered by those subsidiary arrangements, by which the difficulties, necessarily arising from the primary arrangements, are obviated and mitigated ; and by which the greater portion of the earth's surface, has been made accessible to organic beings of the same general character. These subsidiary arrangements it will be our business to explain in the present chapter.

The secondary or subsidiary constituents of climate naturally divide themselves into two great sections ; viz., *those connected with the surface of the earth, as composed of land or water;* and *those connected with the atmosphere.*

In the following sketch of these constituents of climate, we have endeavoured, as usual, to elucidate principles rather than to enter into details; and, as far as is compatible with a general and popular view, have attempted to point out the modes, in which the laws of light and heat, described in the first Book, operate ; so as to produce the phenomena of climate.

SECTION I.

Of the secondary Constituents of Climate, imme-
diately connected with the Surface of the Earth;
and depending on the Nature of that Surface as
composed of Land or Water.

IN attempting to illustrate the operation of the
laws of heat and light in the formation of climate, we shall follow the order, nearly, in which
these laws were discussed in the previous chapters ; that is to say, we shall first consider the
influence of heat and light, as depending on their
latent and decomposed forms ; and afterwards,
their influence as depending on their radiation,
conduction, and convection.

In the prosecution of this difficult inquiry, the
first circumstance which naturally claims our
attention, is the absolute quantity of heat and
light, derived from the sun to the earth.

1. *Of the Proportion of Solar Heat and Light,*
which actually arrives at the Surface of the Earth.
Of the absolute quantity of heat and light derived from the sun to our globe, we have no
means of forming an exact estimate. M. Pouillet
has attempted to show, that the amount of heat
annually received by the earth from the sun, is

equal to that which would be required to melt a stratum of ice nearly forty-six feet thick, and covering its whole surface.* This estimate, however, is to be viewed only as a rude approximation. The difficulty consists, not only in the impracticability of forming precise notions of the heat and light, which actually arrive at any given place in a given time; but in the utter impossibility of forming even a conjecture, of those portions, which become latent, or are otherwise lost, in the passage of the solar rays through the atmosphere. The following observations will give some idea of the absolute quantity of light which reaches the earth ; but it is proper to apprize the reader, that the results stated, are to be considered as liable to much uncertainty. Nor do we know whether they are equally applicable to heat; which, though it obeys laws somewhat analogous to those of light, may nevertheless, have its own peculiar laws.

A vertical ray of light, in its passage through the clearest air, has been calculated to lose at least a fifth part of its intensity, before it reaches the earth's surface. From this cause, and from the actual condition of the atmosphere, it has been estimated, that under the most favourable circumstances; of a thousand rays ema-

* Elémens de Physique expérimentale et de Météorologie, tom. ii. p. 704.

nating from the sun, only 378 on a medium, can penetrate to the surface of the earth at the equator; 228 at the latitude of 45°; and 110 at the poles; while in cloudy weather, these several proportions are a great deal less.*

At present, our attention is solely directed to the portions of heat and light, which thus make their way to the earth's surface. On those portions retained in the atmosphere, we shall offer a few remarks hereafter.

2. *Of the Distribution of Heat and Light over the Earth's Surface in the latent Form.*—The distribution of heat and light in the *latent* state over the surface of the globe, probably follows laws, nearly similar to those of the distribution of sensible heat and light formerly mentioned; that is to say, the quantity latent, like the quantity sensible, diminishes from the equator toward the poles. On this subject, however, we want the necessary data, even for forming an opinion, much less for determining the amount and the exact law of distribution; all of which must be left for future enquirers. But of the infinite importance of the latency of heat, in the economy of nature, the following brief remarks will serve to convey some notion.

Let us take the familiar instance of water;

* Article CLIMATE in the Encyclopædia Britannica. ....

than by which important fluid, the influence of
the latency of heat cannot perhaps be more
strikingly exemplified. We formerly showed,
that the temperature of water in becoming solid
on the one hand, and gaseous on the other;
makes, as it were, a pause; and that these
changes never take place abruptly. The consequence of this arrangement is, that ice and
vapour are formed slowly and gradually; and
as slowly and gradually again become water;
while sudden transitions from one state to the
other are thus entirely prevented. Were it not
for this beautiful provision, we should be constantly liable to inundations, and other incon-
veniencies, which would absolutely have rendered
the world uninhabitable. It is impossible, therefore, to reflect upon the arrangement itself; or
upon the means by which it has been effected;
without being impressed with the most profound
admiration, not only of the wisdom of the Great
Designer of the whole; but of his goodness and
benevolence.

3. *Of the General Distribution of Electricity
and Magnetism over the Earth.*—The recent
discoveries on the connection of electricity
and magnetism, formerly described, have thrown
much light on the distribution of these important agencies over the globe; and the present
extent of our knowledge regarding them, will

be understood by the general reader, from the following summary.

Every one is familiar with the ordinary phenomena of a magnetic needle freely suspended, and with its tendency to assume a position more or less approaching to parallelism to the earth's axis; that is to say; all over the world, a magnetic needle points nearly north and south. Most persons, probably, are also acquainted with the phenomenon termed the *dip* or *inclination* of the magnetic needle: thus, in the latitude of London, a needle exactly poised and freely suspended, instead of assuming a horizontal position, will settle at an angle of 70°, the north pole being downwards. If we carry such a needle southwards, towards the equator, we observe that the dip gradually diminishes; till at a certain point, nearly coinciding with the earth's equator, it has no dip at all, but assumes a perfectly horizontal position. As we still proceed towards the south, the dip again makes its appearance, but in an opposite direction, the south pole being now next the earth's surface. To understand the reason of this *dip* of the magnetic needle and of its general direction, we have only to consider that the earth itself operates as a great magnet, the poles of which are situated beneath its surface. The directive property of the needle is owing to these poles; and when the needle is on the north side of the equator, the north pole of the earth having the greatest effect, the needle is attracted downwards, towards the north pole; hence, exactly over the magnetic pole, the needle would be vertical. Similar phenomena happen in the southern hemisphere; but here the south pole predominates, and, of course, depresses the corresponding pole of the needle; while, at the magnetic equator, from the equal action of both poles, the needle will assume an exactly horizontal position. It may be remarked, that neither the magnetic poles, nor the magnetic equator, coincide exactly with the poles and equator of the earth; and that this non-coincidence is owing to, or rather constitutes, what is termed the *variation* of the needle; which is not only different in different parts of the world; but appears to be liable to periodical differences in the same place, at present not well understood. Such are the principal phenomena of the magnetic needle, as demonstrative of the earth's magnetic operation; we shall attempt to illustrate these phenomena a little further.

We have mentioned, that the earth may be considered as acting like a great magnet. Now, we have formerly shown, that when a magnetic needle is in its natural position of north and south, there exist electrical currents in planes at right angles to the needle, descending on its east side, passing under it from east to west,

P.   R

and ascending on its west side. Hence, we must suppose currents of electricity to circulate within the earth, more especially near its surface, and to be constantly passing from east to west, in planes parallel to the magnetic equator; which electrical currents, if such can be demonstrated to exist, will in their turn completely account for the magnetic directive property of the earth. The next question is, therefore, how far are we justified in assuming the existence of such electric currents within the earth?

We have already alluded to the opinion, that heat occasionally passes into the electric and magnetic energies; an opinion, which, some consider to derive much probability from the phenomena of what has been termed *thermo-electricity;* that is to say, electricity (and magnetism) developed by the unequal distribution of heat through bodies. Now, whether the phenomena of thermo-electricity actually depend on the decomposition of heat, latent or sensible, or upon any other cause, is of little importance; the phenomena themselves are well established; and they seem to account, in the most satisfactory manner, for the general distribution of electricity and magnetism over the earth. The explanation is this: the earth during its diurnal motion on its axis from west to east, has its surface successively exposed to the solar rays in an opposite direction, or from east to west..

The surface of the earth, therefore, particularly between the tropics, will be heated and cooled in succession, from east to west, and currents of electricity, on thermo-electric principles, will at the same time be established in the same direction : now these currents once established from east to west, will, of course, give occasion to the magnetism of the earth from north to south. Hence the magnetic directive power of the earth, in a direction nearly parallel with its axis, is derived from the thermo-electric currents, induced in its equatorial regions by the unequal distribution of heat there present; and depending principally on its diurnal motion.

These recent and beautiful discoveries show, in the most striking manner, that the operations of nature are more extraordinary, and indicate more of simplicity and wisdom of design, in proportion as they are better understood. By what simple expedients, when known, are those wonderful phenomena of the earth's electricity and magnetism produced, which formerly appeared so anomalous and perplexing! And what encouragement do these discoveries hold out to us, respecting future discoveries, which may throw still further light upon the operations of the Great Architect of the universe.

4. *Of the Distribution of Light, in the decomposed Form, over the Globe.*--Every one is fami-

liar with the general fact, that the most splendid
exhibitions of colours of every description, are
displayed in the warmer climates ; and that the
tints of natural objects, generally speaking, become more sad and faded, as we approach the
colder regions; till they merge into the white of
the polar snows. Most persons, also, are aware
of the well known circumstances attending the
total abstraction of light from plants and animals ; and that they thus become more or less
white, or *etiolated.* Hence, we need scarcely do
more than remind the reader, of what must be
already familiar to him, viz., that the decided
colours of tropical productions of every kind ;
whether we consider the gaudy plumage of the
birds ; or the variegated adornment of the fishes
and insects, &c., are so striking, as to be quite
characteristic of these productions. In the
higher latitudes, also, where the contrast between the summer and winter seasons is very
great, the colours of some animals vary with the
seasons; being in the summer generally of some
dark hue, but in the winter nearly white; while
still further, in the polar regions, all is more or
less white ; and the natural covering of the
earth, the snow, is the whitest body in nature.

Putting out of sight the great importance
of the colours of objects, which will fall more
naturally to be spoken of hereafter ; it may be

remarked here, that colours have usually been considered as offering to us, a striking instance of the benevolence of the Deity. Colours are universally agreeable to mankind ; the most incurious and ignorant being attracted by, and delighted with, showy exhibitions of them. Now, all this pleasure is the gratuitous gift of the Creator ; and places his benevolence in the strongest possible point of view. There was no reason why man should have distinguished colours at all, much less have been delighted with them : but what is the fact? not only are we gifted with organs exquisitely sensible to the beauty of colours ; but, as if solely to gratify this feeling, the whole of nature, from the highest to the lowest of her productions, forms one gorgeously coloured picture ; in which every possible tint, is contrasted or associated in every possible manner. Is there a human being who can witness the splendid colouring of the atmosphere above him by the setting sun ; who can witness the beauty and endless variety of tint displayed by every object of the landscape around him, down to the minutest insect or flower or pebble at his feet ; who is conscious of the pleasure he derives from these objects ; and who reflects, that this pleasure was not necessary to his existence, and might have been withheld ? Is there, we ask, a human being who duly considers all these things ; and who will dare to assert, that the Being who made them all is not benevolent?

5. *Of the Laws of Absorption, Radiation, and Reflection of Heat and Light.*—These laws as applied to the earth generally, are at present but very imperfectly understood. The following remarks will serve to convey some idea of the little we know on the subject.

The reader will bear in mind what was formerly stated, that the absorbing power of bodies with respect to heat, and perhaps light also, is *directly* as their radiating power, and *inversely* as their reflecting power. Such is the general opinion ; and, as far as solar heat and light are concerned, this opinion appears to be well founded ; but we shall see presently, that there are strong reasons for suspecting, that the radiating power does not always follow the same law, as the absorbing power. In the mean time, however, we shall proceed to state what has been advanced on these points.

Mr. Daniell has attempted to show, that the absorption, and radiation, of solar heat, increase as we proceed from the equator toward the poles. Thus, in a tropical climate, and under a vertical sun, the greatest extent of the difference between two thermometers, the one covered with black wool, and exposed to the direct rays of the sun, in order that it may absorb to the utmost the incident heat; and the other, uncovered in the shade, is no more than about 47°; while two thermometers, similarly circumstanced, in the middle of summer, in London, give a difference of 65°; and in the Arctic regions, the difference often amounts to 90° at least: so that in the Arctic regions, there is twice as much heat and light absorbed under similar circumstances, as there is in the tropical regions. The same gentleman has also attempted to show (what might have been inferred indeed from the assumed relation between the absorption and radiation of heat and light above mentioned), that the radiation of heat from the earth's surface obeys similar laws; that is to say, that the quantity radiated from the earth, increases from the equator toward the poles. Laws somewhat analogous, and which, when they are better understood, will probably throw much information upon these phenomena, seem to hold with respect to light. Thus we formerly mentioned that when a ray of light falls upon fluids, transparent bodies, or metals, the quantity reflected *increases* with the angle of incidence reckoned from the perpendicular; while the quantity absorbed, of course, *decreases* in the same proportion: but that on the contrary, when a ray falls upon *white opake* bodies, the quantity reflected *decreases* as the angle of incidence increases;

while, of course, the quantity absorbed, *increases* in the like proportion. Hence if heat follow the same law, it is evident that the quantity of heat absorbed by the earth from the solar rays, must *increase* from the equator towards the poles; that is to say, according to the increase of the angle of their incidence, as Mr. Daniell has attempted to show. It is proper, however, to observe that Mr. Daniell's views have been called in question, and that some late observations made in high latitudes do not entirely corroborate them.* We allude to the subject merely with the view of drawing the attention of Meteorologists to it, as one of great interest and curiosity; and as one by no means at present understood. There is every reason to believe, that the absorption (and perhaps the radiation) of heat and light, under some of its modifications, are much influenced by polarization, and consequently by certain angles of incidence and reflection; and that these circumstances, in consequence, have much to do with the distribution of heat and

* We allude here to the observations made in those regions, and given in the appendix to Captain Franklin's Second Journey, by Dr. Richardson, Captain Back, and Lieutenant Kendal. In these observations Dr. R. states that the radiation was much stronger in the spring months, *when the ground was covered with snow*, than in the summer months, when the altitude of the sun was greatest. Dr. R. ascribes this greater radiation to the greater clearness of the air at these seasons; but were there no other reasons?

light, particularly in the higher latitudes ; where they may exert no small influence upon organized beings. The above observations seem to point to the existence of certain general laws, which no doubt hereafter will be elucidated.

In noticing the influence of different colours on the absorption and reflection of heat and light, we stated that black and dark colours generally absorb most and reflect least; and *vice versa*, that white and light colours, reflect most and absorb least. We now proceed to illustrate this interesting subject, by considering the following questions.—Why does whiteness prevail in the Polar regions? Why, for instance, is snow white? On the contrary, why are all sorts of dark and decided colours met with in the tropical climates, except whiteness, which is comparatively rare? Might not snow have been black instead of white; which was just as likely if its colour had been the result of accident? or might not whiteness have been predominant under the equator? Perhaps the best mode of answering these questions, and of placing the subject in a striking view, is to examine what would have been the consequence, if whiteness *had* prevailed under the equator, and blackness at the poles.

As heat and light are supposed to obey nearly the same laws, as far as absorption, radiation, and reflection are concerned ; it is obvious that if white had prevailed in the tropical climates, almost all the solar heat and light, instead of being absorbed, would have been reflected. The consequence of this reflection would have been, that the accumulation of heat, and the glare of light, in the lower regions of the atmosphere, near the surface of the earth, would have been intolerable; and would have rendered these regions quite uninhabitable, at least by the present races of beings. The surface of the earth, also, though it would have been heated slowly, would have been overheated in time; and at length would probably have become so very hot, from its comparatively low radiating powers, that the heat could not have been borne. As it is, from the dark colour of objects near the equator, the heat and light of the sun, there, are readily absorbed, and are as freely given off again by radiation; or perhaps the heat, like the light, is decomposed; and thus the whole is preserved in that comparatively moderate and nicely balanced state, which renders even the hottest parts of the earth's surface inhabitable.

On the other hand, let us consider for a moment what would have been the consequences, if snow had been black; or in other words, if blackness had prevailed in the Polar regions. In this case, all the little light and heat that reach them, would have been absorbed;

and the effect would have been darkness, more or less complete. From the rapid melting also of the snow, on the least exposure to heat and light, we should have been constantly liable to inundations. Thus the whole of the Polar regions of the earth, would have been one dark and dreary void, inaccessible to organic life. But by the present arrangement, all these consequences are obviated. The white snow absorbs a certain portion of light and of heat, (by a beautiful provision *more*, as the angle of incidence increases?) while so much light is reflected as is useful, and no more.* Thus the adjustment of the colours of bodies to the circumstances in which they are placed, constitutes an example of the expedients for obviating those minor incongruities necessarily incidental to the primary distribution of heat and light;

* The reader will observe that, under ordinary circumstances, *white* reflects most, and of course absorbs and radiates *least,* solar heat and light; but if the above remarks on light be well founded, the absorption of light (and heat?) by white bodies *increases* with the angle of incidence. Now, as nothing of this sort is known, or can be well conceived to happen, with respect to radiation, the doubt expressed at the beginning of this section arises, viz., whether, under all circumstances, the radiating and absorbing powers of bodies obey similar laws, even as far as the solar rays are concerned. The absorption and radiation of heat of low intensity, and unaccompanied by light, seem to depend more upon the nature of the surface than upon colour. It must be admitted, however, that at present a great deal of obscurity hangs over the whole of this subject.

and presents altogether one of the most beautiful instances of design, connected with the agency of heat and light.

Lastly, it may be worth while to draw the attention of the reader to the striking contrast displayed between the ponderable and the imponderable forms of matter, as to the ease with which they are decomposed, and the modes in which they exist in nature.

We have seen that to preserve the homogeneity and integrity of ponderable bodies, as of water and air, elaborate arrangements have been adopted, evincing the most extraordinary design and wisdom ; because the decomposition or derangement of water and air, would at once prove destructive to organized beings. But, to preserve the homogeneity of light, and perhaps of heat also, no such care is shown ; because no such care was particularly necessary. The decompositions of these agencies, therefore, are permitted to take their natural course ; and by an admirable provision, so far are colours, &c. from being injurious to us ; that they constitute some of the chief sources of our knowledge and happiness !

6. *Of the Conduction of Heat below the Earth's Surface on Land.*—The soil, from a few inches to a foot or more below the surface, participates

very much in the fluctuations of the surface temperature. In general, perhaps, it may be stated, that the temperature of the surface of the earth, is a little above that of the incumbent atmosphere by day, and below it by night; though much will depend, in this respect, upon the nature of the soil; on its radiating and conducting powers; and on a multiplicity of other conditions, which will readily occur to the reader. At a certain distance, however, below the surface, and varying with the latitude and other circumstances, there must be a determinate stratum, where the temperature is uniform, or nearly so, throughout the year. Experiments on this subject are very limited; but there is reason to believe, that the temperature of this *invariable stratum* coincides nearly, with the mean annual temperature of the place; and that its depth below the surface, in different latitudes, varies between forty and eighty feet. The reader need scarcely be reminded, that the well known uniformity of the temperature of cellars and caves, depends chiefly upon the circumstances we are now considering. As an instance of the uniformity of temperature in such places, it may be mentioned, that a thermometer placed in the caves under the observatory in Paris, at a depth of about eighty-five feet below the surface, has, during fifty years, scarcely varied more than a quarter of a degree from 11·82° of the centigrade scale; equal very nearly to 53¼° of Fahrenheit.

A few experiments have been made to determine the variation of the temperature, throughout the year, at different depths from the surface, down to the invariable stratum; and the following is a summary of the results, which, perhaps, may be considered as generally applicable to the northern hemisphere.

In the month of August, the temperature of the earth decreases, in nearly a uniform manner, from a little below the surface, to the stratum of invariable temperature. In the month of September, the temperature is nearly uniform to fifteen or twenty feet below the surface; beyond which depth, the temperature decreases a little and slowly, to the stratum of invariable temperature. During the months of October, and November, the temperature increases from the surface, to the depth of fifteen or twenty feet; and below this depth, it remains nearly uniform to the invariable stratum. During December, January, and February, the temperature, being at its minimum upon the surface, increases in a manner nearly uniform, downwards to the invariable stratum. During March, and April, there is a· rapid decrease of temperature to the depth of one or two feet; below this depth, the temperature decreases less rapidly;

and still lower, the temperature increases a little. During the months of May, June, and July, the temperature being at its maximum, at the surface, decreases downwards, but less rapidly and to a greater depth ; it then begins to increase a little, till it attains the temperature of the invariable stratum. The rapidity and degree, however, with which these changes take place, as well as the changes themselves, appear to fluctuate very considerably, not only in different places under the same Isothermal line, but in the same place in different seasons.

Since heat is propagated through the soil by conduction, of course it is propagated in all directions. Hence, it may be supposed to move laterally as well as downwards ; and, generally speaking, the temperatures of contiguous spots probably tend to equalize each other. But upon the whole, the influence of the lateral propagation of heat through the solid parts of the earth must be very limited.

7. *Of the Propagation of Heat and Light below the Earth's Surface in Water.* Water is a very imperfect conductor of heat, in the usual acceptation of the term. Thus, almost any degree of heat may be applied, for a considerable time, to the upper surface of a mass of water, without materially influencing the temperature below ; so imperfectly and slowly is heat

conducted through this fluid. The process by which heat is communicated through water, we have termed *convection*. When heat is applied to the bottom of a vessel full of water, the portion of the water first heated, expands in bulk, and thus becomes specifically lighter; it then rises to the top, carrying with it the newly acquired temperature; while another cold portion, sinking to the bottom, is heated in turn; and so on, till the whole mass becomes uniformly heated.

With respect to the propagation of light through water, it has been calculated, that not a tenth part of the incident light, can advance five fathoms downwards in the most translucent water; that even of vertical rays, one half is lost in the first seventeen feet; and that these vertical rays become reduced to one-fourth by traversing thirty-four feet, which correspond to the mass of an atmosphere. It thus follows, that only the hundred thousandth part of the vertical rays, can penetrate below forty-seven fathoms; which is scarcely equal to the glimmer of twilight; and that the depths of the ocean must be always in perpetual darkness.*

Such are the general principles according to which heat and light are propagated in water. But in speaking of this fluid in a former chapter, we alluded to one of the physical properties of

* Article CLIMATE, in the Encyclopædia Britannica.

water, of the utmost importance in the economy of nature, and which, perhaps, almost more than any thing else, indicates design; since, like the composition of the atmosphere, this property of water constitutes an exception, as it were, to a general law, expressly directed to a particular object. We have mentioned that it is a general law, that all bodies, in every state of aggregation, expand by heat and contract by cold: now water forms a marked exception to this law. Like other bodies, water continues to contract on the removal of heat, till its temperature comes down to within a certain distance (7° or 8°) from its freezing point. At this distance, water begins again to expand, and the expansion continues till it becomes ice; at which moment of freezing, a sudden and considerable expansion takes place. Hence, the specific gravity of ice, is decidedly less than that of water; and the solid necessarily swims on the surface of the fluid. The importance of this anomalous property of water, is so great, that it is doubtful whether the present order of nature could have existed without it; even although every thing else in the world had remained the same. For instance, were it not for the comparative lightness of ice; this solid, instead of beginning to be formed at the *surface* of water, would have begun to be formed at the *bottom;* as the colder water, from its greater specific gravity, would

P.   S

naturally have sunk : for similar reasons, also, the lowest stratum of ice would have been the last to have melted. Now, let us reflect for a moment upon the consequences of such an arrangement. In the northern, and even in temperate, climates, the bottoms of all lakes and deep waters would have been a mass of ice, and totally inaccessible, therefore, to organized beings. During the summer, a few feet of the upper part of the ice would, perhaps, have been melted ; but what little had thus been melted in summer, would again have become solid during winter ; and as the accumulations of ice would have been constant ; all the seas, even perhaps to the tropical climates, at least at their bottom, would, long before this time, have been a mass of ice ! But what in reality happens ? In consequence of the above anomalous properties of water, this mischief is entirely prevented ; and not a particle of ice can be formed in a lake or other collection of water, till the whole mass is cooled down to the temperature of 40° ; at which temperature, the specific gravity of water is at its maximum.

These properties of water operate in the following manner. On the application of cold to the surface of water, the cooled portion sinks, and its descent forces up a portion of warmer water to the surface, which after communi-

cating some of its heat to the superincumbent
air, sinks in its turn ; and this process goes on
for a greater or less time according to the depth
of the water. If the depth be not very considerable, the whole body of the water becomes
cooled down to 40°; at which temperature the
specific gravity not increasing, the circulation
ceases ; and the *surface* of the water, (not the
bottom) becomes at length, so far cooled, as to be
covered with ice. If the depth of the water be
considerable, the application of cold may be
long continued without the result of freezing;
hence, in this, and in other, countries, not intensely cold, it often happens, that deep lakes
remain unfrozen during the coldest winters.

The above anomalous properties of the expansion of water and its consequences, have always
struck us as presenting the most remarkable
instance of design in the whole order of nature—
an instance of something done expressly, and
almost (could we indeed conceive such a thing
of the Deity), at second thought, to accomplish
a particular object. Further, if in conjunction
with this anomalous property of water, we take
into account the still more anomalous constitution of atmospheric air, and at the same time
consider the relations of water and air to organic
existence ; we are unavoidably driven to the
conclusion, that the Maker of water and of air,
has designedly created these anomalies, to

obviate difficulties which would have rendered organic existence a physical impossibility. Nor do the suppositions which the sceptic will urge, that these properties of water and air flow naturally from their constitution, diminish the force of the argument. The force of the argument lies, in the first place, in the fact, that water and air have been created with such anomalous properties; and, in the next and chief place, *that these anomalous properties have been brought into action precisely where they are required.* Moreover, the argument is greatly strengthened, by the fact that *two anomalies,* rather than that *two ordinary circumstances,* have been thus expressly adjusted.

Having stated the general principles on which heat is distributed through water, and its most remarkable consequence; we are now to enter into a few details with respect to some other consequences of this distribution. Of these, one of the most striking is, that the temperature of the water, at the bottoms of deep lakes, or inland seas, must remain nearly uniform during the whole year. Thus it has been found, that the temperature of the water at the bottoms of many of the lakes in Switzerland, often varies no more than 3° or 4°; while the temperature of the surface, often fluctuates 20° or 30°. Hence in deep waters, in temperate climates, the changes of temperature are chiefly confined to the upper strata of the water; nor can ice (except from some very sudden and powerful accessions of frost) form on the surface of such a lake, till, as before observed, the whole of the water in it, is cooled down to 40°; at which temperature all circulation ceases. When a coat of ice has been once formed; this ice, as we shall see presently, has also a powerful tendency to prevent the further cooling of the inferior strata.

With respect to *waters in motion*, as small streams, or rivers of no great depth and magnitude, and containing fresh water; though unfavourably circumstanced for freezing, they do nevertheless congeal. The process usually commences at the shores, where the water is shallowest, and its motion is least rapid; from whence, the ice gradually advances towards the middle of the stream. When the whole of the surface has once become fixed; congelation goes on actively, particularly by night. As the thickness of the ice increases, however; the quantity added daily, even supposing the cold to remain the same, gradually diminishes; on account of the bad conducting power of the ice. Hence, in a block of ice taken from a river or lake, we may often observe the strata corresponding with the daily, or rather nightly additions, presenting a gradually decreasing series, from several inches, down to a few lines in thickness.

Of the Temperature of the Waters of the Ocean

at great Depths.—Between the Tropics, the temperature of the ocean diminishes with the depth; in the Polar seas, on the contrary, the temperature augments with the depth. In the temperate seas, comprised between 30° and 70° of latitude, the temperature of the water gradually decreases as the latitude increases, until about the latitude of 70°; when the temperature begins to rise, as before mentioned. Hence, about the latitude of 70°, there exists a zone or band, at which the mean temperature of the ocean is very nearly constant at all depths. The temperatures of particular parts of the ocean, however, have been observed to be much influenced by the depth and extent of the water; particularly in high latitudes.

We have already mentioned the influence of the saline matters of the ocean upon the *freezing point of sea-water;* and we have now to point out the important consequence of this property, in the economy of nature. In its natural state, sea-water freezes at about 28° or 29°; but when it has been concentrated by previous freezing, the congealing point is reduced to 15° or 16°: while water saturated with salt, it is said, does not freeze at a temperature above 5°. Besides this property of lowering the freezing point of sea-water, the saline matters also increase its specific gravity; and affect its point of maximum density. Hence, from these circumstances, and from their immense depth and extent, the waters of the ocean resist freezing, still more effectually than even running fresh water; and are indeed rarely frozen, except in latitudes where the most intense and unremitting cold prevails.

Of the under Currents of the Ocean, existing between the Equatorial and Polar Regions.— That the diminished temperature of the waters of the ocean, at great depths near the equator, could not have been acquired in the torrid zone, is evident; nor, on the other hand, could the comparatively high temperature of the waters, at the bottom of the Polar seas, have been acquired in the frigid zone; at least this high temperature of the Polar seas, cannot be caused from without. Hence it has been supposed, that there is a constant interchange going on between the waters of the Equatorial, and those of the Polar regions; though there are considerable difficulties, at present, as to the means by which this interchange is effected. These difficulties arise principally from some uncertainty, with respect to the point of maximum density of sea-water; which does not appear to be satisfactorily established. Whether in the profound, and comparatively quiescent abyss of the ocean, the process of diffusion, or the central heat of the earth formerly alluded to, exert any influence, we have no means of determining. But if a central heat really do exist; its effects must be considerable, particularly within the frigid zone. Whatever be the cause of this approach to uniformity of temperature throughout the waters of the ocean at great depths, all over the globe ; its use in the economy of nature, in tending to equalize the distribution of temperature, cannot be questioned ; since it constitutes one of those beautiful provisions, by which the difficulties of the distribution of temperature, necessarily incidental to the earth's figure and motions, are obviated : whilst among the minor circumstances contributing to the same end, may be mentioned the tides, and the innumerable superficial currents produced by winds, and by other causes, which are to be considered elsewhere.

8. *Of the Differences of Temperature, as depending upon whether the Surface be Land or Sea.*—When speaking of the distribution of temperature over the earth's surface, we alluded to the differences between insular and continental climates; and perhaps it may not be amiss, to make a few remarks upon the actual general amount of the differences of temperature, as produced by land and water.

In the middle of oceans, and far from the influence of land, the diurnal change of temperature of the air near the surface of the sea, is

much less than upon land. Thus, in the equatorial regions, the greatest difference between
the temperature of the day and that of the night,
at sea, is said to amount to 3° or 4° only ; while
upon land the difference often amounts to 9° or
10°. In temperate regions, and particularly in
latitudes extending from 25° to 50°, the difference between the maximum and the minimum
diurnal range of the thermometer, at sea, is still
very trifling, amounting only to 4° or 6° ; while
upon the continents, as for example, 'at Paris,
the range often amounts to 20° or 30°. To these
circumstances it is owing, that small insular
situations, partaking of the character of the
surrounding ocean, are much less liable to great
diurnal changes than continents ; and hence, in
general, they possess more equable climates.

   Both by sea and land, the minimum temperature takes place about sunrise. The maximum temperature at sea occurs about noon, or
very soon after ; while upon land, it takes place
from two to three hours after noon. Between
the tropics, the maximum temperature of the air
is said to exceed a little that of the surface of
the sea. But when the temperatures are observed at short intervals, as for example, every
four hours, and all the temperatures are compared, the results are different ; and they seem
to show, that even between the tropics, the temperature of the surface of the sea is higher than

that of the incumbent atmosphere. Between the latitudes of 25° and 50°, the air is rarely warmer than the surface of the sea; and in the Polar regions, it is very unusual to find the air as warm as the sea; it is in fact almost always colder, and generally very much colder.

As connected with this part of our subject, it may perhaps, before we close, be desirable to offer a few remarks upon *the temperature of natural springs*, and their relation to the mean temperature of the earth, at the places where they make their appearance.

Springs discharging large quantities of water, and thus indicating that they come from considerable depths below the surface of the earth, preserve nearly the same temperature during the whole year. In our hemisphere, what little augmentation of temperature springs undergo, is generally in the month of September, while they are coldest in the month of March; though the differences seldom exceed two or three degrees. If we compare the temperature of the springs of any place, with the mean annual temperature of that place; we find that there is a near connection between the two, all over the globe. In the torrid zone, however, the mean annual temperature of the air is usually higher by three or four degrees than that of the springs;

while in the temperate zone, on the contrary, the springs are warmer than the air. The excess of temperature of springs, as compared with the mean annual temperature, goes on increasing with the latitude; so that, between 60° and 70° of latitude, this excess amounts to from 5° to 7°. Other things being the same, the temperature of springs varies considerably, according to their copiousness; as a large body of water, will be less liable to be influenced by the surrounding soil, than a smaller body of water; and may even, in turn, influence the temperature of the soil itself.

The subject of *thermal springs*, as intimately connected with the history of volcanoes, belongs to the Geologist.

We have thus enumerated the principal circumstances connected with the distribution of temperature upon the surface of the earth, and at such parts below the surface as are within our reach. We now come to the second great division of the subject of climates; viz., climate as depending on the atmosphere.

SECTION II.

Of the Secondary Constituents of Climate imme-diately connected with the Atmosphere.

THE phenomena of the atmosphere, originally constituted the proper study of the Meteorologist, and even yet they claim the largest share of his attention. The subject, in all its bearings, is very extensive, and many of the details are imperfectly understood. We shall endeavour to present a brief outline of the principal phenomena under the following heads.—*Of the distribution of heat and of light through the atmosphere, and of the consequences;—of the distribution of water through the atmosphere, and of the phenomena dependent on that distribution;* and, lastly, *—of the occasional presence of foreign bodies in the atmosphere.*

1. *Of the Distribution of Heat and of Light through the Atmosphere, and of the Consequences.* —Every one is familiar with the general fact of the diminished temperature of the higher regions of our atmosphere; and that in the hottest countries, by ascending a lofty mountain, we encounter, at different heights, every variety of temperature, even to the temperature of per-

petual snow, and of the Polar regions. One of the first circumstances, therefore, which claim the attention of the Meteorologist, is the law of the distribution of sensible heat, or of temperature, through the atmosphere.

The law of the distribution of temperature through the atmosphere is tolerably uniform : though it is occasionally liable to variations and interruptions, depending upon local differences ; and perhaps upon other circumstances not satisfactorily understood. The mean results of a great number of observations, made in different parts of the world, appear to show, that for every 100 yards of altitude, Fahrenheit's thermometer sinks one degree. This statement, probably, does not, within moderate limits, differ much from the truth ; though some late researches have rendered it probable, that while at different heights the rate of the decrease of temperature is uniform, the rate of altitude increases constantly, and according to laws very similar all over the world ; that is to say, supposing the first 252 feet are equal to one degree ; the second degree will be equal to 255 feet ; the third to 258 ; the fourth to 261 ; &c.

- The causes upon which this great cold of the higher regions depends, are chiefly the two following ; first, the perfect permeability of the atmosphere to the solar rays ; on which account, they radiate through it almost without affecting its temperatures, till reaching the earth, they exert their utmost influence; and secondly, the increased capacity for heat possessed by air, in proportion as it becomes more rare. From the first of these causes it happens, that the temperature of the lower regions of the atmosphere is derived, not immediately from the sun, but from the earth. The surface of the earth absorbing the solar heat, recommunicates it to the immediately incumbent atmosphere; while all the higher portions of the atmosphere remain unaffected. For though, from diminished specific gravity, heated air naturally ascends, yet as its capacity for heat at the same time increases, ascending air rapidly loses its sensible heat; as in the second place we have to explain.

Dr. Dalton, and afterwards Sir John Leslie more completely, have attempted to show, that the equilibrium of heat in an atmosphere is obtained, when *each of its molecules*, or in other words, when *the same weight of air*, in the same perpendicular column, is possessed of *the same quantity of heat*. Now, since atmospheric pressure diminishes with the height according to a certain law; it is obvious, that the *same weight of air*, at the surface of the earth, and in the higher regions, will occupy *very different spaces*. But since the absolute quantity of heat is exactly the same in both portions; it is likewise obvious, that in the higher regions of the atmosphere, from the increased capacity of the air for heat, the quantity of latent heat is augmented, while the quantity remaining sensible, becomes less. Hence the temperature of the air diminishes as we ascend, exactly in the proportion that its latent heat, that is to say, its capacity for heat as produced by rarefaction, increases. In consequence of this arrangement, to use the words of Dr. Thomson, " if a quantity of cold air were suddenly transported from an elevated region to the surface of the sea, its density would be continually increasing during its descent, while its latent heat would diminish in the same proportion ; and when it reached the level of the sea, its temperature would be just as high as that of other portions of air in the same latitude and elevation. Air, therefore, does not feel cold in consequence of falling from an elevated situation, though this be an opinion commonly entertained ; but in consequence of its being suddenly transported from a more northerly, to a more southerly situation."* Thus, to the above beautiful and simple law, we owe the permanent state of equilibrium of temperature in the atmosphere ; which, in spite of all the disturbances constantly produced by minor causes, from the natural tendency to right itself, is never very seriously affected.

* On heat and electricity, p. 129.

Of the Limits of Perpetual Snow.—Connected with the diminution of temperature in the higher regions of the atmosphere, are the *limits of perpetual snow* in different latitudes. These limits, of course, may be naturally supposed to follow the mean temperature of 32°, from the level of the sea in the Polar regions, to the highest point of their range under the equator. This inference is obvious, and, generally speaking, correct; though it is liable to certain modifications, and to some anomalies, of which the following are the most remarkable.

Under the equator, the limits of perpetual snow are the most fixed and steady, and seem to exist generally at an altitude of between 15,000 and 16,000 feet. As we recede from the equator, the oscillations for the most part become more striking, and all the phenomena assume a more irregular form. Such, for example, is the case in the Mexican Cordilleras; but still more evidently in the Himmala range; where there is a difference of no less than 4,000 feet, between the limits of perpetual snow on the northern, and on the southern sides of the mountain; that on the northern being the highest. As we proceed toward the temperate zones, we find, in mountainous countries, below the limits of perpetual snow, immense bodies of ice, or *glaciers*, as they are termed. These glaciers are formed by the alternate melting and congealing of the exten-

sive beds of snow that lie above them. The glaciers, accumulating in valleys, are often by the
enormous and increasing weight of the snow
and ice in the upper parts, pressed downwards
far beyond the limits of the snow itself. Such
are the *glaciers* of Switzerland, of Norway, and
of other countries in temperate climates. All
these circumstances, with others which might be
mentioned, and many probably that are unknown
to us, combine to render the limits of perpetual
snow irregular. These irregularities are so considerable, that Humboldt, from numerous observations, has inferred the limits of perpetual
snow at the equator to be nearly 3° above the
freezing point ; while in the temperate zone,
they are nearly 5° below that point ; and in the
frigid zone, no less than 10° or 11° below freezing ; results which seem to prove, that the general temperature of the air decreases in the equatorial, otherwise than in the colder regions.
From the peculiar distribution of the land in the
southern hemisphere, little is known of the line
of perpetual snow in that part of the world ; but
it will probably be found to be different from
the line in the north, and generally lower.

The perpetual snow resting on the tops of
mountains, constitutes a most important provision in the economy of nature, particularly in
the warmer climates ; where the accumulated
snow becomes the prolific source of innumerable

P.   T

rivers, without which those regions would be uninhabitable.

In the accompanying map, we have endeavoured to show the analogy between the effects produced on the distribution of temperature, &c. by height above the surface of the earth, as compared with difference of latitude. There is, however, one striking difference between high and low situations, which must have considerable influence upon organization, though this influence has not been studied so carefully as it ought to be; viz. the difference of atmospheric pressure. At the surface of the earth, the atmospheric pressure is nearly the same in all latitudes; but as we ascend above the surface, the pressure rapidly diminishes. Every thing else, therefore, being supposed to be the same, the difference of pressure will probably render certain provisions and accommodations necessary, of which, at present, we are ignorant; but which might doubtless be much elucidated by a careful study of Alpine plants and animals, as compared with those occupying the plains. Another circumstance, which must materially influence organization, is the great intensity of light in the mountainous districts of tropical climates, as compared with the intensity of light at the surface of the earth, in the corresponding climates, in high latitudes. The diminished intensity of light, however, in high latitudes, is doubtless compensated for, in some degree, by the greater length of the day.

Of the Distribution of Heat and Light through the Atmosphere in their latent Forms. —In the preceding paragraphs, we have alluded to the quantity of heat existing *latent* in the higher regions of the atmosphere. But besides this quantity, which may be supposed to be common to the whole atmosphere ; the distribution of latent heat and light must, in some degree, follow the same law as that of sensible heat and light ; that is, must decrease from the equator toward the poles. Thus there can be no doubt, that the expanded air of the equatorial regions, contains much more heat and light in the latent state, than the comparatively dense and dry atmospheric air of the Polar regions ; and it is probable that the rigours of each extreme are mitigated by this provision. The distribution of electricity through the atmosphere, seems also to be regulated by very similar laws. It may, however, be remarked, that the effects of heat and light, in the latent form, as well as the effects of electricity, are much more striking as connected with the water in the atmosphere, than with the constituents of the atmosphere itself. We shall, therefore, defer what we have to say on those subjects, till we speak of the water in the atmosphere.

Of the Propagation of Sensible Heat through

the Atmosphere.—The propagation of sensible heat through the atmosphere is chiefly effected by the process termed *convection.* Convection, of course, implies motion, or currents; which currents, as existing in the atmosphere, we need scarcely observe, are denominated *Winds.* The winds, therefore, are of the utmost importance in the economy of nature, as tending to equalize the distribution of temperature over the globe; and the following brief explanation will serve to give a general knowledge of their nature.

Atmospheric currents may be considered under two heads: those of a general kind, extending more or less over the whole globe; and those which are produced by various transient derangements of the distribution of temperature, the effects of which are limited to particular localities. On each of these we shall make a few remarks.

The *general currents* of the atmosphere depend principally upon the two following circumstances, which, if borne in mind, will furnish the reader with a clue to the whole subject: viz. the unequal temperature of the equator and of the poles; and the diurnal motion of the earth upon its axis. The convective operation of the first of these general causes may be thus illustrated. We have stated that the entire pressure of the atmosphere all over the earth's surface is nearly the same, and equal to that of a column of mercury about thirty inches in height. We have also stated that the mean temperature of the atmosphere near the equator, and at the level of the sea, is upwards of 80°; while in the Polar regions, it is constantly below 32°, the freezing point of water. Hence, as air expands by heat, and becomes specifically lighter; it is obvious that a given bulk of air at the level of the sea round the poles, must be considerably heavier, than a similar bulk of air at the level of the sea under the equator. The air, therefore, round the poles being colder and heavier, will have a tendency to flow along the earth's surface from the poles towards the equator, and to displace the lighter air under the equator; while the equatorial air so displaced, will, owing to its lightness, ascend and flow back again over the colder air, north and south toward the poles, so as to preserve the equilibrium. Moreover these currents will be perpetual; for the heat of the equator and the cold of the poles being constant, the same tendency to change will always exist, and thus the currents will be constant likewise.

These atmospheric currents constitute one primary element of the winds; and are the grand means by which the equalization of temperature over the globe is effected. If the earth were at rest, and free from irregularity, the currents or winds near its surface would, of course, be in the northern hemisphere, always due north; and in the southern hemisphere, due south; while the velocity would, in each case, gradually diminish from the poles towards the equator; where there would be a perpetual calm.

But the earth is in a constant state of motion upon its axis from west to east; by which motion, the currents are deflected from their northern and southern course towards the east. This eastern deflection constitutes the other primary element of the winds, to be next considered.

The general reader will bear in mind; that on the surface of a globe, revolving like the earth on its axis, the motion of any given point at the equator, is the *greatest*, and at the poles, the *least* possible. Thus while the poles are quiescent, the velocity of any given place at the equator of our earth, is about 1000 miles an hour; from which extreme, the velocity gradually diminishes toward the poles. This motion of the earth on its axis, operates in the production of an easterly current in the atmosphere, as follows. Supposing there were no atmospheric currents from the north and south towards the equator, and that the earth revolved upon its axis as at present; one of two things must happen. Either the earth during its revolution would carry with it the incumbent atmosphere; in which case there would be a perpetual calm over its surface: or the earth would revolve within the atmo-

sphere, leaving, as it were, the atmosphere behind it; in which case there would be an apparent current or wind over the whole of the earth's
surface, in a direction opposite to that of the
earth's motion, that is from east to west; which
wind, supposing the atmosphere did not move
with the earth, would, of course, be at its maximum at the equator. Now both these causes
are continually operating, and give origin to all
the variety of the eastern currents upon the
earth's surface; which, with the northern and
southern currents, before described, conspire to
produce the well known currents, called *the trade
winds*. Before we attempt to explain the trade
winds, their phenomena may be thus, briefly
described.

The trade winds in the Atlantic ocean, extend
to about 28° on each side of the equator. At
their extreme northern and southern boundaries,
these winds generally blow from the east: but
as they proceed towards the equator from the
north and from the south, they gradually pass
from the east, through all the intermediate points
of the compass; till near the equator, they become in the northern hemisphere, due north;
and in the southern hemisphere, due south.
The trade winds are subject to some slight
variations, chiefly arising from the position of
the earth with respect to the sun. On these
variations we do not think it necessary to en-

large. The general phenomena are as stated; and the principles advanced, appear to offer the following explanation of these phenomena:

In the temperate regions of the earth, the winds seem to obey no certain laws; at least, no laws so determinate as those of the trade winds. But about the tropics, both in the northern and in the southern hemispheres, the operation of the double currents and motions before described, becomes distinctly perceptible. Thus about the tropics, the surface of the earth begins to move faster than the incumbent atmosphere; and hence in these regions, the prevailing currents are from the east. Indeed near the tropics, the currents are nearly due east; principally on account of the great and somewhat sudden change of temperature produced by the vertical sun of the tropical regions; which may be supposed to interfere with, and perhaps to check momentarily, the regular progress of the great northern and southern currents. As we proceed, however, towards the equator; the atmosphere, in both hemispheres, gradually acquires the velocity of the earth; while the intensity of the eastern current, diminishes in the same proportion, and at length entirely disappears. At the same time, the currents from the north and the south continuing, slowly deflect the currents, from the east towards the north, in the northern hemisphere; and from the east towards the south, in the southern hemisphere; till left alone by themselves, the polar currents proceed onward to the equator, as if the motion of the earth had no existence.*

To these great atmospheric currents may be traced the fluctuations of the barometer, and all the innumerable modifications peculiar to different localities of sea and land, of mountain and plain. For, as Mr. Daniell justly observes, in the nicely balanced state of the forces producing these currents, slight irregularities of temperature are capable of causing great disturbances; and expansions and contractions acting unequally upon the antagonist currents, operate by deranging the adjustment of their several velocities. Hence accumulations in some parts, and corresponding deficiencies in others, necessarily arise; and occasion fluctuations in the barometer, far surpassing what would be occasioned by the whole vapour, supposing it were at once

---

* In the first edition of this volume, we ascribed the above theory of the trade winds to Mr. Daniell. We have since learnt, however, that the same theory was advanced a century ago, by Mr. Hadley. (Philos. Trans. xxxix. p. 58.) Without being aware of the existence of Mr. Hadley's essay, Dr. Dalton offered a similar explanation of the trade winds in his Meteorological Observations and Essays. Still more lately, the same theory was advanced by Mr. Daniell, and illustrated by Captain Basil Hall, in an essay appended to the second edition of Mr. Daniell's Meteorological Essays; to which we refer the reader for details.

added, or annihilated. At the same time, these irregular distributions, in struggling to restore the equilibrium, produce temporary and variable winds, which modify the regular currents, and often reverse their courses, particularly in the temperate regions; where, as formerly mentioned, the alternations of temperature, and the fluctuations of the Barometer, are the most remarkable.

Such are the elements of the general currents pervading our atmosphere; and such the modes in which these currents modify extreme temperatures and their consequences. The same causes are constantly operating in different forms and degrees; so as to produce all the infinite variety among the winds, which we observe in nature. These are so numerous and diversified, as actually to baffle all attempts at explanation or arrangement; we shall therefore content ourselves with one instance only, by way of illustration, viz., *the sea and land breezes.*

The explanation of what are denominated the sea and land breezes is very obvious; and is not less applicable to many similar phenomena. During the day, the surface of the land acquiring heat, imparts its temperature to the incumbent air. This air expanding in bulk becomes specifically lighter, and rises in consequence; while the cooler air from the surrounding sea rushes in to supply its place, and thus

produces the current called the *sea breeze.*
During the night, on the contrary, the waters
of the ocean part with their heat much more
slowly than the land, and the reverse action, or
the *land breeze* takes place. In hot climates
near the sea shore, and in insular situations,
these alternations constitute a most agreeable
variety.

2. *Of the Presence of Water in the Atmosphere.*—In the foregoing section, we have endeavoured to give an outline of those beautiful
provisions, which by means of the air of the
atmosphere, have been adopted to prevent the
consequences necessarily arising from the unequal distribution of heat and light over the
globe. We now come to another subject of not
less importance, viz., the phenomena depending
on the existence of water in the atmosphere;
and which, taken together, principally constitute
what we emphatically denominate *the Weather.*

The phenomena depending on the existence
of water in the atmosphere may be considered
under the four following points of view:—First,
*Of the phenomena of evaporation and condensa-
tion; and of the general dependance of vapour
on temperature:*—Secondly, *Of the conditions of
an atmosphere of vapour alone; and of a mixed
atmosphere of vapour and air:*—Thirdly, *Of the
general relations of evaporation and condensation,*

*as they exist in our atmosphere; and of the cir-
cumstances by which these relations are influenced:
—*Fourthly, *Of the distribution of heat and light
in their latent and decomposed forms, through the
vapour of the atmosphere; and of the effects of
that distribution.*

First, *Of the phenomena of evaporation and condensation; and of the general dependance of vapour
on temperature.*—We have before stated the fact,
that water assumes the elastic form, in a greater
or less degree, at all temperatures. From the
tendency of water, thus to rise " above the Fir-
mament;" not only the ocean, but ice and
snow, are unceasingly contributing their supply
of moisture to the air; and this important fluid,
so indispensable to vegetable and animal existence, is distributed over the surface of the whole
earth. In considering, therefore, the relations
of the water of the atmosphere to temperature;
the phenomena which first claim our attention,
are the processes by which water is taken up,
and again separated, from the atmosphere; that
is to say, the processes of *Evaporation* and *Condensation.*

In treating of the nature of *Evaporation*, the
questions to be answered at the outset are,—
Why is moisture present in the atmosphere?
By what force is its presence determined, and its
quantity limited? The reply to these questions
depends upon the properties of matter in general,
and of vapour in particular, as formerly des-

cribed. These properties, if borne in mind by the reader, will enable him to understand what follows.

When water is exposed to the air in an open vessel, the molecules of its uppermost or superficial stratum being released from the influence of those below them, have a natural tendency to assume that degree of polarity which is appropriate to their temperature. Hence, after acquiring the latent heat necessary to produce this polarity, either at the expense of a portion of their own sensible heat, or of that of the atmosphere; the superficial molecules of water become self-repulsive, and fly off into space in the form of vapour. If the space over the water be circumscribed and be a vacuum; the molecules fly off with such rapidity as instantaneously to fill it. But, if the space be occupied by air, or be of indefinite magnitude; the molecules fly off more slowly, so as gradually to diffuse themselves through the whole space; quite on the same principle, and in the same manner, that one gaseous body is diffused through another.

Such, in few words, may be deemed a simple statement of what evaporation is. We shall next proceed to enquire into the nature and operation of the means by which evaporation not only takes place, but is limited within certain boundaries.

In a former chapter, we remarked, that the

elastic force exerted by all bodies in the gaseous
state, bears a certain relation to their temperature ; but that the degree of this elastic force
varies according to other circumstances ; particularly, according to whether the gaseous body,
at the given temperature, be capable of existing
in the fluid or in the solid states, as well as in
the gaseous state. Thus, atmospheric air, not
only at the temperature of 32°, but at all known
temperatures, is a gaseous body ; and, under
ordinary circumstances, exerts an elastic force
equal to the weight of a column of mercury 30
inches high : whereas, at the same temperature
of 32°, water is a solid ; and the force of the elasticity of its vapour, is not more than equal to
about 1-5th of an inch of mercury. But at, and
above 212°, its boiling point, water, under ordinary circumstances, can exist only as a gas;
and in this gaseous form, and at the temperature
of 212°, water obeys precisely the same laws,
and exerts the same elastic force, as atmospheric
air would obey and exert, under similar circumstances. Hence it will be readily understood,
that the law of the elastic force of vapour below
212°, is very different from the law of that force
above 212° ; as by experiment is found to be
the fact.

From the preceding remarks it will appear,
that all other things being the same, the tendency of water to assume the form of vapour,

or the rate of its evaporation, as well as the actual quantity of water in the state of vapour in the atmosphere, will increase as the temperature increases. We need not state in detail, the exact law of this increase. It is sufficient for our purpose to observe, that at all temperatures below the boiling point of water, that is to say, at all common atmospheric temperatures; while the rate of the increase of temperature is *slow* and uniform, or in an *arithmetical* progression; the corresponding rate of the elastic force of vapour, by which the quantity of water as vapour is determined, increases much more *rapidly*, or nearly in a *geometrical* progression. This important fact is connected with several most interesting circumstances.

The phenomena of the *Condensation* of vapour from the atmosphere, are next to be explained. As the quantity of water in solution in the atmosphere, can never be greater, though it may be less, than the quantity proper to the temperature; when vapour, or what is the same thing, when a portion of air saturated with vapour, at any given temperature, is cooled below the point of saturation; a portion of the vapour is separated in the form of fluid water, while the remainder assumes the elastic condition proper to the newly acquired and diminished temperature. The forms assumed by the water so separated, are various; and depend very much upon the quantity separated; and on the separation taking place in atmospheric air. When the quantity of water separated is small; the minute detached particles diffused through a large space, are suspended in the atmosphere by its buoyancy, and assume the form of what, for the sake of distinction, we shall call *Visible Vapour*, viz. mists, clouds, &c. When the quantity separated is greater; the particles collect into drops too large to be upheld by atmospheric buoyancy, and they fall to the earth in the shape of rain, hail, &c.

Of the two great processes of evaporation and condensation, it may be further remarked, that by a beautiful provision, they have a constant tendency to limit each its own operations: evaporation is increased by heat, and produces cold; condensation is produced by cold, and liberates heat. Moreover, in virtue of another wonderful arrangement; by evaporation, water is separated entirely from all foreign bodies, and is thus condensed in a state of absolute purity.

Secondly, *Of the conditions of an atmosphere of vapour alone; and of a mixed atmosphere of vapour and air.*—We now come more particularly to consider the mode in which vapour exists in the atmosphere. To facilitate the understanding of the subject, we shall commence by supposing the air to be absent; and shall enquire what would be the conditions of *an atmosphere of vapour* under the pressure and temperature existing at the surface of the earth, and at different heights above the earth's surface.

As the elastic force of vapour increases faster than the temperature of the vapour; and as the mean temperature at the Equator is, at least, 80°, while at the Poles it is below 32°; it follows, that in an atmosphere of vapour, heated similarly to that of our earth, the specific gravity of the vapour at the Equator, would greatly exceed the specific gravity of the vapour at the Poles. Vapour thus exhibits a condition directly opposite to that of air, under the same circumstances. Hence the tendencies of the lateral currents, in an atmosphere of vapour, at the surface of the earth, would be precisely the reverse of those in an atmosphere of air; the tendency of the currents would be from the Equator toward the Poles; instead, as in air, from the Poles toward the Equator.

We have elsewhere stated the law of the decrease of the temperature of the atmosphere, observed in ascending from the surface of the earth; the atmospheric air being supposed to be free from moisture. A similar law would regulate the decrease of temperature in an atmosphere of vapour; but the rate of decrease would be much more slow, than in an atmosphere of perfectly dry air. Thus under the Equator,

P.   U

where, at the level of the sea, the mean temperature is at least 80°; the temperature of an
atmosphere of perfectly dry air would sink to
the freezing point at a height of 15,000 feet:
while the temperature of an atmosphere of vapour would, at the same height, sink only to
70°. At all the parallels of lower mean temperature, onward to the lowest round the Poles,
at any height above the level of the sea, similar
differences would exist between the temperature of an atmosphere of perfectly dry air, and
the temperature of an atmosphere of vapour;
these differences, of course, varying with the
mean surface temperature. At the same time,
throughout the whole range, from the Equator
to the Poles, the specific gravity of the vapour at
the level of the sea, would always exceed its
specific gravity at any height above. Hence,
in an atmosphere of vapour, there would be no
tendency to vertical currents.

Having thus stated the leading properties of
an atmosphere of air, and of an atmosphere of
vapour, separately, we come to the proper subject
of our inquiry; viz., *the condition of an atmosphere resulting from a mixture of air and vapour*
—of such an atmosphere, indeed, as that in
which we actually live.

The reader will have no difficulty in understanding the nature of a mixed atmosphere;
provided he has clearly apprehended what has

been above stated, regarding the simple atmo-
spheres which are its components; and will advert to two other circumstances, that are now to
be noticed. These two circumstances are intimately connected with the principles previously
stated, and with each other; and an exposition
of them is absolutely necessary for obtaining
a true knowledge of the relations of an atmosphere of vapour, with an atmosphere of air.
These circumstances have not been mentioned
sooner; the consideration of them having been
intentionally delayed, in order that their influence might be seen, where their application
is more immediately requisite. They are as
follow:

When vapour and air are mixed together, the
resulting volume of the mixture depends on the
amount of the elastic forces of the vapour and
of the air; not on any relation between their
volumes. Thus when a cubic foot of air at the
temperature of 32°, and exerting an elastic force
equal to 30 inches of mercury, is mixed with
a cubic foot of vapour, having the same tempe
rature, and exerting an elastic force equal to
only 1-5th of an inch of mercury; the volume of
the mixture resulting, is not two cubic feet, but
only 1·0066 foot. Hence, as the addition of
vapour to air adds comparatively little to the
bulk of the air, and consequently diminishes
only in a trifling degree its specific gravity;

the great aerial currents formerly described as pervading the atmosphere, are scarcely affected by the vapour they contain.

When two portions of vapour, having different temperatures, are mingled together; or when a portion of vapour is brought into a state of mixture or contact, with a portion of water, or with any other body colder than the vapour; *the resulting mean temperature*, whatever that may be, is, in both cases, the temperature which regulates the elastic force of the mixture. Now, since the elastic force of vapour increases most rapidly from the temperature of 32° to 212°, the increase being in a geometrical progression; while the increase of the temperature is in an arithmetical progression; it follows, that when two portions of vapour, of equal bulk, but of different temperatures, are mixed together; or when a portion of vapour is brought into contact with any solid colder body; the resulting mean temperature is always *below* that requisite to preserve the water in a state of vapour. Hence, such mixture or contact is always followed by a portion of the vapour being condensed into water. In a future part of this section, it will be necessary to illustrate further this important fact; but a familiar instance may be noticed here. Let us suppose that a pound of water at the temperature of 212°, which being in a state of steam would occupy a space of about 27 cubic feet, were suddenly brought into mixture with a pound of water at a temperature of 32° ; the effect would be an instantaneous condensation of the greater part of the steam into water. For, the resulting mean temperature would obviously be far short of 212°, below which temperature, the elastic force of vapour most rapidly diminishes. On this property of vapour, depends the working of the common steam-engine.

The reader is thus at length prepared to enter on the complicated subject of a mixed atmosphere of vapour and of air.

We have shown, that the rate of decrease of the temperature of an atmosphere of vapour, in ascending from the earth's surface, would be very much slower than that of an atmosphere of air. Now, at all temperatures, the existence of atmospheric air is permanent; while the very existence of vapour is dependent on temperature : it follows, therefore, that in a mixed atmosphere of vapour and of air, the quantity of vapour contained in the mixture, is regulated solely by the temperature of the air : that is to say, the quantity of vapour present in an aerial atmosphere, can never exceed, though it may be less than, the quantity which is proper to the temperature of the air. If the quantity of vapour in such a mixed atmosphere, be precisely the quantity that is proper to the temperature of the air; such an atmosphere is said to be *saturated* with vapour.

But, neither at the earth's surface, nor at any height above it, can the degree of saturation of a mixed atmosphere of air and vapour, be quite equal to that which is proper to the temperature of the air ; and the difference between these two degrees of saturation, augments from above downwards. The cause of this difference may be thus explained. The rate of increase of the temperature of air, from above downward, being in arithmetical progression; and the air being, in a mixed atmosphere, that ingredient which controls the whole mixture : the rate of increase of the tension of the vapour, instead of following the geometrical rate which belongs to it as vapour; is obliged to conform to the arithmetical rate of increase of the temperature of the air. The result of this controlment necessarily is, that the quantity of vapour present in a mixed atmosphere will, at any successive diminution of the height above the surface of the earth, become successively less and less than that which would be required to saturate the air. An example will make this result evident.

At the Equator, as we have said, the temperature of the air, about 15,000 feet above the level of the sea, is nearly 32°. Now, for the sake of illustration, let us suppose the air at this height to be saturated with vapour. From Dr.

Dalton's table of the tension, or elastic forces of vapour at different temperatures, it appears that the tension of vapour at 32° is equal to the weight of ·200 inch of mercury ; and that the difference between the tension of vapour at 32° and the tension of vapour at 33°, the value, namely, of the first term or unit, in our assumed arithmetical series, is ·007 inch of mercury. Now, the difference between 32°, and 80°, the mean temperature at the level of the sea under the Equator, is 48°. Supposing, therefore, each of these 48 degrees to increase in an arithmetical progression, ·007 for each degree; the tension for the whole 48 degrees will amount to ·336 ; which tension added to ·200, the tension at 32°, gives ·536 inch, as the tension corresponding to the vapour at 80°, the temperature of the earth's surface under the Equator. But, by Dr. Dalton's same table of tensions, we find that ·536 does not represent the proper tension of vapour at 80°, but of vapour at about 61° only. According to this estimate it follows, that at the Equator, while the temperature of the air over the earth's surface is 80°, the point of saturation with vapour is 19° below that temperature. Hence, at the Equator, the air immediately incumbent on the earth's surface must be comparatively dry. Moreover, the cause which has been thus shown to produce the dryness of the Equatorial air, at the earth's surface, must all over the globe exert different degrees of the same influence. *The air, everywhere incumbent on the earth's surface, must, therefore, always be under the point of saturation;*—the relative degree of dryness being highest under the Equator, and gradually diminishing as we recede north or south toward the Poles.*

In such a mixed atmosphere as we have supposed, and as in reality surrounds our globe, if its equilibrium be undisturbed, and if it be conceived to be at rest; the admixed vapour will have nearly the same tendencies to motion, which would exist in an atmosphere of pure vapour, formerly described. But, from the more equal distribution of the vapour, when mingled with air; the contrasts between the specific gravities of different portions of vapour, in different parts of the atmosphere, will be much less striking, than if the atmosphere consisted of vapour alone. Consequently, the rates of motion depending upon such differences of specific gravity, will be less remarkable in a mixed atmosphere,

* The mathematical reader will observe, that the quantities given in the text are not rigidly accurate, but are intended only for familiar illustration of the principles regulating moisture. The truth is, as has been noticed in the text, in *no part* of a vertical column of a mixed atmosphere, in a condition of equilibrium and at rest, can the air be in a state of saturation. It has been remarked, that the degree of saturation often continues nearly uniform up to a certain point, and then suddenly decreases.

even though saturated with vapour, than they would be in a purely aqueous atmosphere; while in an unsaturated atmosphere, the motions of the vapour must be still more liable to be influenced by the motions of the air, than they would be in an atmosphere of air, at its utmost point of saturation.

Before we close this part of our subject, let us reflect for a moment, on the consequences of such a state of comparative dryness of the lower atmosphere next the earth. Over the greater portion of the earth, the air which, during the day at least, is warmed by contact with the earth's surface, and thus becomes lighter, has, as we have observed, a constant tendency to rise into the higher atmosphere. Now, if this air were saturated with vapour; of course, whenever the air by rising became mixed with colder air, its vapour would be more or less condensed, and a cloud would be formed. Hence, if we lived in such an atmosphere; we should be always enveloped in a mist, through which the sun would not be visible. But, by the benevolent arrangement we enjoy, this consequence is so entirely prevented, that, unless under peculiar circumstances, and always for beneficial purposes, the air at the earth's surface is hardly ever saturated with moisture. The air which has been warmed by contact with the earth, can, therefore, rise from the surface, without any condensation of its moisture within the limits of its point of saturation. Thus, at the Equator, before the air reaches the temperature of 61°, the presumed point of its saturation, it must ascend to the height of 6000 or 7000 feet. At this height, its vapour will be condensed, and a cloud will be formed; which may either be precipitated on the spot from which its constituent vapour had risen; or may be transported by the currents of the atmosphere, similarly to refresh a distant country; or may be again dissolved in the air: while under all these contingencies, the whole of the lower portion of the atmosphere is exempt from mist, and continues perfectly transparent. These operations are unceasing; moreover, the very clouds, by giving out their latent heat, and shielding the earth's surface from the direct influence of the sun, produce a still further effect; and have a constant tendency to modify their own formation and existence.

The general result of all the complicated and beautiful machinery connected with the properties of vapour, is, as before observed, that water is constantly ascending into the atmosphere, where it is again condensed in the form of rain, &c. over the whole earth. We shall therefore examine a little more in detail, the relations of the two great processes of evaporation and condensation, by which these important arrangements are accomplished.

Thirdly. *Of the general relations of evapo-ration and condensation as they exist in our atmo-sphere; and of the circumstances by which these relations are influenced.*—We have already de-scribed the general phenomena of evaporation and condensation, and have stated the laws on which these phenomena depend. It will, therefore, in this place, be sufficient to remind the reader, that the degree, and the rate, of evaporation, though they increase with the temperature, are regulated chiefly by the ex-isting degree of saturation of the air. That is to say, under all temperatures, evaporation decreases, as the air which receives the vapour, approaches its point of saturation. Hence it follows, that in an atmosphere perfectly sa-turated with moisture, and in a state of ther-mal and dynamical equilibrium, there can be neither evaporation nor condensation. The processes of evaporation and condensation, there-fore, always indicate a disturbance of the thermal equilibrium in some part of the atmosphere : condensation denoting a depression of the tem-perature below the mean, or point of thermal equilibrium : evaporation, on the contrary, de-noting that the temperature in some part of the atmosphere has been raised above the mean ; or at least that the temperature having been depressed below the mean, is again undergoing an elevation to the mean point. Evaporation and condensation may be thus considered as mutually dependent; so that one process cannot take place without the other. For this reason, in the great expanse of nature, these two processes oscillate or fluctuate .about the point of equilibrium, within certain limits which are never passed; and which limits, though subject to countless anomalies, in general, decrease from the Equator toward the Poles.

With respect to the temperature which constitutes the point of equilibrium; in an atmosphere of vapour, that point would, of course, be the maximum point of saturation. But in a mixed atmosphere of vapour and air, like that of our globe, the point of equilibrium cannot be the point of utmost saturation, but must be that *inferior point of saturation* formerly described, as being determined by the temperature of the predominant air. Thus at the Equator, where the mean temperature at the level of the sea is about 80°, the mean point of saturation will, according to our former estimate, be 61°; while in London, where the mean annual temperature is about 49½°, the mean point of saturation, (or the *dew point*, as it is termed,) has been fixed by Mr. Daniell at 44½°. In temperate climates, the mean point of saturation, at any particular place, varies with the seasons from day to day, being higher in summer than in winter. During any shorter period, as that of a day and night, the mean point of saturation, as might be expected, generally bears a certain relation to the *lowest* degree to which the temperature has fallen during the period; since the *Hygrometer** shows that the degree of saturation, at any hour, is seldom below the point of saturation corresponding to the lowest temperature of the twenty-four hours; at which point it continues nearly uniform, so that the point of saturation during the warmer parts of the day generally varies only a few degrees. The elevation and depression of the dew point in temperate climates is thus another, and unceasing cause of change; and produces a variety in evaporation and condensation so great, as to baffle any attempt at accurate enquiry.

From what has been said, it will appear that in a mixed atmosphere, the rate of evaporation and of condensation, other things being equal, will depend, not on the difference of the temperature of the air from the maximum point of saturation, but on the difference of the temperature of the air from that of *the mean dew point;*

* The Hygrometer is an instrument for measuring the degree of moisture of the atmosphere. Mr. Daniell's Hygrometer is here alluded to, which is the only one acting upon scientific principles. *Daniell's hygrometer* shows the degree of temperature at which water is deposited from the atmosphere, and consequently its state of saturation.

that is to say, will increase or diminish as this difference increases.

Of the nature and causes of the motions of the vapour through the air of the atmosphere.—Water is dispersed through the atmosphere in two ways: , viz. by the motions of vapour properly so called; and by the motions of visible vapour or clouds.

With respect to the motions of *vapour properly so called;* it may be remarked, that, vapour is circulated through the atmosphere, partly by convection; but chiefly, by means of that tendency before described, which water possesses at all temperatures, to assume the form of vapour, and to diffuse itself through the air. In a mixed atmosphere of vapour and air, the motions of the vapour, on the large scale of the operations of nature, are considerably influenced, no doubt, by the motions of the air. For example, large masses of air, more or less saturated with vapour, in proportion to their respective temperatures, and having either vertical or lateral motion, must carry with them the vapour they contain; whether there be much or little vapour so contained. On the other hand, motions of the air, on a smaller scale, as we shall presently see, may be even caused—may certainly be accelerated or retarded, according as the diffusive motion of the vapour, to be next considered,

may agree with, or may be opposed to, these
motions of the air.

In an atmosphere of vapour, when the temperature, and consequently the elasticity, of any
portion is reduced; the surrounding vapour, by
virtue of its greater elastic force, continues to
advance towards the cooler locality, and to be
there condensed until the thermal equilibrium is
restored. The motion thus arising, depends
upon the dynamical properties of vapour; and
in an atmosphere of vapour, this restoration of
the dynamical equilibrium, which depends on
the thermal equilibrium, would take place with
so great rapidity, as to be almost instantaneous.
But in a mixed atmosphere, the case is different:
in such an atmosphere, the presence of the
heavier and more abundant air modifies, in a
remarkable degree, the rapid motion of the
lighter and less abundant vapour. Hence, instead of a rush of vapour and a momentary
deluge, the diffusive motions of the vapour take
place slowly; and sudden evaporation and condensation, with their consequences, are effectually prevented. These tendencies to diffusive
motion, in vapour of different temperatures,
have, no doubt, great influence on the contiguous surfaces of large masses of air differently
saturated; and, in particular, are liable to affect
smaller masses of air differently saturated, when

they are in the immediate neighbourhood of each other. Thus, as we have already noticed, the disturbance of the equilibrium of the vapour, may be to such an extent, in some portion of the mixed atmosphere; that the surrounding vapour, urged to move by its tendency to restore the equilibrium, may occasionally be supposed to drag with it the air and the clouds, and thus to produce local currents. For instance, let us imagine a mass of warm and almost perfectly dry air, to be brought into the neighbourhood of another mass of air of precisely the same temperature, but saturated with vapour. The two masses of air, from being of the same temperature, would, as air, have no tendency to intermingle. But as being portions of a mixed atmosphere of vapour and air, the dryer air would be, as it were, a vacuum, towards which the vapour from the moist air would have a tendency to flow, till both masses of air became equally moist. In such a case, the motion of the vapour might be supposed to cause more or less of motion in the air, while a momentary cloud would probably be formed ; which cloud would be dissipated when the equilibrium was restored. In this way, it is likely that some of the minor motions of the atmosphere are produced.

The *motions of visible rapour*, or clouds, are

principally caused by currents in the atmosphere, or winds; and by electricity. Although the proper motions of vapour are liable, as we have said, to be considerably influenced by atmospheric currents; yet vapour is much less liable to be so affected, than visible vapour; for when once the vapour in the atmosphere has been separated, and has assumed the form of *visible* vapour, its own proper powers of motion cease; and it becomes entirely subject to those of convection. Visible vapours, therefore, of all kinds, from their being liable to be wafted by every breeze, are in a constant state of motion; and are thus frequently carried where vapour, in virtue of its mere diffusive property, would never reach.

Another undoubted cause of the motions of visible vapour or clouds, is Electricity. In the form of vapour, electricity is not known to exert any influence on the motions of water through the atmosphere; but the moment the water is precipitated, and assumes the form of visible vapour, it becomes subject to electrical attractions and repulsions, by which the motions of clouds, &c. are much liable to be influenced; as will be more particularly explained hereafter.

Of the accidental circumstances affecting evaporation. — The accidental circumstances which principally operate to affect the rate of evapora-

P.   X

tion, are the greater or less extent of the evaporating surface; and the velocity and degree of saturation of the current of air over that surface. But besides these causes of variation, there are other circumstances which probably have great influence on evaporation ; some of which are to us of the utmost interest, as being brought more immediately in contact, as it were, with our existence. The chief of these additional circumstances affecting evaporation which we shall notice are ; — circumstances incidental to the water which undergoes evaporation ; and circumstances incidental to the air into which the water is evaporated.

In speaking of the circumstances incidental to water, we may remark, in the first place, that the condition of the water as ice, does not, so far as is known, affect in the least degree the rate of evaporation, as might be expected. Thus Howard mentions an instance in the month of January, in a certain year, when the vapour, from a circular area of snow five inches in diameter, amounted to 150 grains between sunset and sunrise ; and before the next evening, 50 grains more were added to the amount, the gauge having been exposed to a smart breeze on the housetop. Under like circumstances an acre of snow would, in the course of twenty-four hours, evaporate the enormous quantity of 64,000,000 grains of moisture ! Even by the evaporation during the night only, a thousand gallons of water would, in that short time, be raised from an acre of snow. It may thus be easily understood, how a moderate fall of snow may entirely vanish during a succeeding northerly gale, without the slightest perceptible liquefaction on the surface.* We have given this statement to satisfy the general reader of the fact, that evaporation is constantly going on from snow and ice; indeed there is every reason to believe, as before observed, that the quantity thus evaporated is precisely equal to what would be evaporated, from water itself; provided that body could exist as a fluid at the same temperature.

The circumstances incidental to water, and affecting evaporation and saturation, arise chiefly from its purity or impurity. The presence of foreign bodies, as of saline matters, for instance, is well known to raise considerably the boiling point of water; in other words, they lower its tendency to become vapour, and thus diminish its evaporating and saturating powers. Hence the air over the sea, though, of course, much nearer, in general, to the point of saturation appropriate to the latitude and temperature, than air over the land, is comparatively seldom in a state of perfect saturation; and sea-water, so far

* Article Meteorology; in the Encyclopædia Metropolitana.

from being capable of saturating the air with
moisture up to the dew point, has even the
power of abstracting a portion of the moisture
from an atmosphere so saturated, and of thus,
to a certain extent, drying the air.

Evaporation on land is precisely similar to
evaporation from sea-water; since the various
rocks and soils may be considered as so many
saline matters, diminishing, in their several
degrees, the tendency to become vapour possessed by the water united with them. Hence,
under like circumstances, some rocks and soils
are dry, while others are moist; so that, in proportion to the evaporating powers of the rocks
and soil of a country, will that country be liable
to all the consequences of dryness or of dampness of soil. Plants also seem to differ much in
their capacity for retaining water. The dryness
of a country will, therefore, be considerably
affected by the nature of its vegetation; and
the predominance of certain plants or trees in
a district may thus increase the dampness of
its soil.

Regarding the effect which *foreign matters in
the atmosphere* have in influencing evaporation
from the subjacent land or water; we are unable
to speak with as much confidence, as we have
spoken of the controlling power of the foreign
matters in the water itself. There is, however,
reason to believe that certain constituent prin-

ciples of the atmosphere are occasionally associated, so as to form compounds; and that these
compounds, acting as foreign bodies, materially
influence evaporation. See Appendix.

*Of the accidental circumstances which influence
condensation.*—The condensation of vapour from
the atmosphere, as we have already stated, differs in some degree, according to the origin of
that diminished temperature by which the condensation is produced. We shall, therefore,
commence with those phenomena of the precipitation of moisture from the atmosphere, which
depend on the *radiation* of heat from the earth's
surface into space. The most remarkable of
these phenomena are *Dew, Hoar Frost,* and
certain forms of *Mist.*

Of Dew.—The phenomena of dew were first
satisfactorily explained by the late Dr. Wells;
who showed by the most decisive experiments,
that, apparently, all these phenomena were
owing to the effects of the radiation of heat from
the earth's surface, during the absence of the
sun. The reader is referred to Dr. Wells'
" *Essay on Dew,*" for details. It is sufficient for
our present purpose to observe, that when the
direct influence of the sun is removed in the
evening, and the surface of the earth thus no
longer continues to acquire heat; at that instant,
from the ceaseless activity of heat to maintain a
state of equilibrium, the surface of the earth,

being the warmer body, radiates a portion of its superfluous temperature into the surrounding space; and thus the air immediately in contact with the surface, becomes cooled below the point of saturation, and gives off a portion of its water in the form of dew.

We formerly stated, that the radiating powers of bodies differ exceedingly according to their composition, the nature of their surface, their colour, &c. These differences, of course, produce corresponding effects on the deposition of dew; and, as beautifully demonstrated by Dr. Wells, explain its greater or less deposition under certain circumstances, or its entire absence under others. Thus, what formerly appeared so extraordinary, viz. why in the self-same state of the atmosphere, &c. one portion of the earth's surface, or one portion of herbage, should be covered with dew, while another in the immediate neighbourhood should remain dry, is no longer a mystery; but is perfectly explicable on the supposition of their different radiating powers.

The deposition of dew is always most abundant during calm and cloudless nights, and in situations freely exposed to the atmosphere. Whatever interferes in any way with the process of radiation, as might be expected, has a great effect on the deposition of dew. Hence, the radiation of heat, and consequently the deposition of dew, are obviated, not only by the slightest covering or shelter, as by thin matting, or even muslin; by the neighbourhood of buildings, and innumerable other impediments, near the earth's surface: but matters interposed at a.great distance from the earth's surface have precisely the same effect. Thus clouds effectually prevent the radiation of heat from the earth's surface; so that cloudy nights are always warmer than those which are clear, and in consequence, there is usually on such nights little or no deposition of dew.

From dew there is an insensible transition to *Hoar Frost;* hoar frost being in fact only frozen dew, and indicative of greater cold. We observe, therefore, that frosty nights, like simply dewy nights, are generally still and clear.

The influence of radiation in producing cold at the earth's surface, would scarcely be believed by inattentive observers. Often on a calm night, the temperature of a grass plot is 10° or 15° less than that of the air a few feet above it. Hence, as Mr. Daniell has remarked, vegetables, in our climate, are, during ten months of the year, liable to be exposed at night to a freezing temperature; and even in July and August to a temperature only two or three degrees warmer. Yet, notwithstanding these vicissitudes, in the words of the same gentleman,—" To vegetables growing in climates for which they are originally designed by nature, there can be no doubt that the action of radiation is particularly beneficial, from the deposition of moisture which it determines upon the foliage; and it is only to tender plants, artificially trained to resist the rigours of an unnatural situation, that this extra degree of cold proves injurious."* It may be observed, also, that trees of lofty growth frequently escape being injured by frost, when plants nearer the ground are quite destroyed.

Of Mists and Fogs.—Mists are not necessarily connected with the deposition of dew; because during the deposition of dew, the atmosphere often continues transparent, even to the earth's surface. At other times, however, and for reasons which, in the present state of our meteorological knowledge, cannot be satisfactorily explained; the deposition of dew is accompanied by a visible vapour or mist, more or less dense, and extending from the surface of the earth to a greater or less height in the atmosphere. When mists, from other causes, are general, and extend to considerable heights above the earth's surface, they acquire the name of *fogs*. The optical properties, and the buoyancy in the atmosphere, of mists and fogs, would seem to indicate that they are not formed of solid particles, but of minute hollow

* Meteorological Essays and Observations, p. 511, second edition.

vesicles, having the quality of mutual repulsion ; the tendency to repel each other, preventing the coherence of the vesicles into drops, at least under ordinary circumstances. These vesicles have been occasionally observed of considerable magnitude. Thus Saussure, in one of his Alpine journeys, saw vesicles float slowly before him having greater diameters than peas, and whose coating seemed inconceivably thin. It is proper to mention, however, that there is diversity of opinion respecting the actual constitution of visible vapour.

That the cause of the formation of mists and of fogs is, to a certain extent, similar to that of the formation of dew, appears by their prevalence over rivers and large masses of water, especially during the autumnal months. The radiation of heat from the land, and from the water, is at these seasons very different; the difference being greatest, when the temperature of the water approaches 40°, its point of maximum density. The water is then of a temperature nearly uniform, both by day and by night ; while the temperature of the land is, during the day, much higher than 40°, and during the night, often much under that temperature. The water in most cases occupying the lowest situations ; whenever, from the inequalities of the surface of the land, or from any other cause, the colder air produced by radiation over the land, is made to mix itself with the warmer air over the water; the moisture in the warmer air is condensed, so as to become mist. Hence the formation of mist differs slightly from that of dew, inasmuch as there is occasionally (not always) an intermixture of air of different temperatures. The reason is thus evident of the fogs and mists so frequently seen over rivers and in valleys ; or in other situations, where there is a collection of water. The occurrence of these mists is usually on clear and cold nights,—oftener in autumn, and seldom or never in cloudy weather; the state of the atmosphere having exactly the same influence on these mists, as on the deposition of dew. There cannot be a doubt that these mists, like clouds, produce a great effect in impeding radiation, and in thus mitigating the intensity of cold. Mists are therefore of much importance in the economy of nature. Plants growing in low grounds are by them shielded from the destroying influence of the sudden cold, which would almost certainly be produced, not only by the free radiation of heat in such situations, but by the descent of cold air from the surrounding high grounds.

The fogs which hang over great towns admit of an explanation similar to that of other aqueous fogs. The air of the town being warmer than the air of the surrounding country, and being at the same time charged with moisture nearly to the point of saturation, is, in cold weather, suddenly cooled, either by the radiation of its own heat, or by the admixture of the neighbouring cold air ; while the superfluous moisture is condensed as a fog.

The fogs of high latitudes, more especially the fogs of the Polar seas, are in the same manner owing to the radiation of heat. The cooling of the warmer air, over the immense masses of floating ice, gives rise to an unequal distribution of temperature, and thus at certain seasons, to uninterrupted fogs. It is probable that in all these instances, the fogs beneficially alleviate the severity of cold, by checking great and sudden alternations of temperature ; which would otherwise interfere much with the operations of organic life.

Fogs have been sometimes observed of a strong odour, apparently the result of an admixture of foreign bodies. In a subsequent paragraph these fogs will be fully considered.

Of Clouds.—From mists and fogs the transition to clouds is easy and natural ; as clouds, in reality, are nothing more or less than masses of visible vapour, precisely similar to that composing fogs, but existing at a distance above the earth's surface. Clouds differ principally from mists and fogs in their mode of formation. Thus mists, like dew, as we have seen, are the result of the cooling of the lower strata of the atmosphere by radiation. Fogs are so far the result of radiation, that they usually arise from the influence, which air cooled by radiation, exerts on warmer air.   While clouds probably depend altogether on *convection;* and result from the intermixture of strata of air of different temperatures, and in different states of saturation, in the higher regions of the atmosphere.

Such is the general opinion of the formation of clouds : but it must be confessed that there are considerable difficulties about the subject; and that the mere assumption of strata of different temperatures, more or. less saturated with vapour, and having the motions supposed to depend upon such different temperatures and degrees of saturation, seems quite inadequate to account for all the phenomena connected with the formation and appearance of clouds.

From the principles formerly stated when we described the phenomena and properties of a mixed atmosphere of air and vapour, it appears, that clouds in general must be formed at that elevation in the atmosphere, in which the mean temperature of the air becomes equal to, or falls below the point of saturation of such air.   This elevation, which may be said to constitute the *region of clouds,* must of course be highest under the Equator—an inference supported by fact; for it has been observed that within the tropics, the clouds are most frequently higher than in the temperate zones ; and in the temperate zones, the clouds appear to be higher in summer than they are in winter. In the temperate zones, Gay Lussac thinks that clouds, in general, are upheld at an average distance from the earth's surface of between 1500 and 2000 yards. Occasionally, however, clouds have a much greater altitude ; and the *Cirrus*, a form of cloud to be presently described, has been seen far above the greatest elevation hitherto attained by man.

In some parts of the world, clouds are rarely seen ; while in other parts, the sky is seldom cloudless. Such extremes are usually confined to extreme climates, or depend upon local causes. In the temperate zones, from the irregularity of the atmospheric currents, and from the other innumerable circumstances calculated to disturb the equilibrium of the atmosphere, the general character of clouds varies much, even under the same parallel of latitude. Hence all the infinite variety of sunshine, of cloud, and of shower, which more especially distinguish the temperate zones, and our own variable sky in particular ; where they exert such constant and commanding influence upon our comfort and wellbeing, as to become almost interwoven with our very existence.

Though clouds are of such endless diversity of *figure* and appearance, they have been classed by Howard under three primary forms, and four modifications. The three primary forms are :

The *Cirrus*, composed of fibrous-like stripes, parallel, flexuous, or diverging, and extensible in all directions.

The *Cumulus*, heaped together in convex, or in conical masses, and increasing upwards from a horizontal base.

The *Stratus*, spreading horizontally in a continuous layer, and increasing from below.

The first of these forms, the cirrus, is confined chiefly to the higher regions of the atmosphere. The second form, the cumulus, occupies a lower but still an elevated station; while the third form, the stratus, usually rests on the surface of the earth, constituting the mists already described in this chapter.

Of the four modified forms of clouds, two are intermediate, and two are composite.

The first of the intermediate forms is the *Cirro-Cumulus*, consisting of small roundish, and well-defined masses in close horizontal arrangement.

The masses that compose the second intermediate form of clouds, the *Cirro-Stratus*, are likewise small and rounded, but are attenuated towards a part, or towards the whole, of their circumference. They are sometimes separate: when in groups, their arrangement is either horizontal, or slightly inclined, and the masses are either bent downwards, or are undulated.

Of the two composite forms of clouds, the first

is the *Cumulo-Stratus*, made up of the Cirro-Stratus blended with the Cumulus; the Cirro-Stratus being either intermingled with the larger masses of the Cumulus, or widely enlarging the cumulous base.

The second composite form, and the last of the four modifications, of clouds, is the *Cumulo-Cirro-Stratus*, or *Nimbus*, the *rain-cloud;* being that cloud, or system of clouds, from which rain is falling. The nimbus is a horizontal layer of aqueous vapour, over which, clouds of the cirrous form are spread; while other clouds of the cumulous form, enter it laterally and from beneath.

A little attention will enable any one to discriminate these varieties of clouds; at least when their forms are well defined. Yet, it must be acknowledged that clouds often assume forms to which it is difficult to give a name.

With respect to the *motion* of clouds, it may be remarked that there is not perhaps a more frequent subject of optical delusion, nor any thing regarding which, we are more liable to be mistaken. Into such enquiry it would be quite inconsistent with the design of this treatise were we to enter minutely; but we offer the following brief illustration. Let us suppose a cloud moving from the distant horizon towards the place where we stand. Let us also suppose that the cloud, during its motion, retains its size and figure unchanged; and that it proceeds along its course in a uniform horizontal line. A cloud so moving, when first seen, will appear to be in contact with the distant horizon; and will thus necessarily, from its remote position, appear to be much smaller than in reality it is. During its advance towards us, the cloud will seem to rise into the sky, and to become gradually larger, till it is almost directly overhead. Continuing its progress, it will then seem again to descend from the zenith, and to lessen in size as gradually as it had before increased; till at last it vanishes in the distance, opposite to where it commenced its movement. Thus the same cloud, without deviating from its motion in a straight line, and retaining throughout the same size and figure, will, by optical delusion, seem continually to vary in magnitude. The line of its motion also, instead of being straight, will appear to be a curve having its vertex directly above us, and its extremes boundless in opposite points of the horizon. We have given the most simple case that can be supposed. But clouds, as they exist in nature, are unceasingly varying in shape, in magnitude, in direction, and in velocity; so that to form a just estimate of their figure and direction, or to unravel their motions, becomes absolutely impossible.

After what has been stated, it will be superfluous to dwell upon the *uses* of clouds in the economy of nature; we shall therefore briefly remind the reader of a few only of the most obvious benefits derived from clouds. The first of these that claims our attention, is, that on the large scale at least, clouds constitute a sort of intermediate state of existence between vapour and water ; by which, sudden depositions of water, and their consequences, are entirely prevented. If all the water separated from the atmosphere fell at once to the earth, in the state of water, we should be constantly liable to deluges and other inconveniences ; the whole of which are obviated, by the present beautiful arrangement. Again, clouds are one great means by which water is transported from seas and oceans, to be deposited far inland, where water otherwise would never reach. Clouds also greatly mitigate the extremes of temperature. By day, they shield vegetation from the scorching influence of the solar heat, and produce all the agreeable vicissitude of shade and sunshine : by night, the earth, wrapt in its mantle of clouds, is enabled to retain that heat which would otherwise radiate into space ; and is thus protected from the opposite influence of the nocturnal cold. These benefits arising from clouds are most felt in countries without the Tropics, which are most liable to extremes of temperature. Indeed, clouds constitute one great means by which, in temperate climates, the extremes of heat and cold are regulated. Lastly, whether

P.   Y

we contemplate them with respect to their form, their colour, their numerous modifications, or, more than all, their incessant state of change; clouds prove a source of never-failing interest, and may be classed among the most beautiful objects in nature.

Having finished the consideration of the various states of visible vapour; we are now to examine the phenomena of the precipitation of water from the atmosphere in the form of *Snow, Sleet, Rain,* and *Hail.* We shall first speak

Of Snow.—We commence with snow, because it offers the most simple case of the precipitation of water from the atmosphere ; snow being nothing more than the frozen visible vapour composing clouds. Hence a flake of snow, examined with a high magnifier, exhibits a beautiful display of minute crystals, often possessing the greatest variety of form.

When the temperature of the atmosphere, down to the earth's surface, is constantly below the freezing point ; it is obvious that any moisture separated from the atmosphere must assume the solid form. If the quantity separated be small, the frozen particles of water remaining detached, float in the atmosphere in the form of crystallized spiculæ, and thus give origin to what is called the *frost-smoke;* a phenomenon not unfrequently witnessed in polar latitudes. Even in temperate climates, the same thing has been supposed occasionally to take place in the higher regions of the atmosphere, and thus to produce certain optical phenomena, to which we shall hereafter refer.

The above are comparatively rare phenomena. Most generally, the quantity of water separated, is so large, that the crystallized particles are agglutinated together into masses or *flakes*, and thus fall to the earth in the form of snow. When the quantity deposited is very great, as is often the case; there can be no doubt, that the causes operating to produce such large deposition, are precisely similar to those which produce rain in warmer climates ; and which will be considered in a subsequent paragraph.

Such, in few words, are the principles upon which snow is formed ; and from these, the reason is at once apparent, why during the winter in temperate climates, and throughout the whole year in the Polar climates, most of the water that falls to the earth assumes the form of snow.

We formerly mentioned how much we owe to the *whiteness* of snow ; and we may now remark that we owe still more, to its *low conducting properties*, and to its *lightness*. Thus, by its low conducting properties, snow shields vegetation from the rigorous cold of the higher latitudes ; where every thing herbaceous would be destroyed during the winter, were it not for the protecting influence of snow. Again, if the water which now descends to the earth as snow, were to be precipitated in the form of solid masses of ice; vegetation would be destroyed, and the whole of the colder parts of the earth would be uninhabitable!

It has been remarked, in temperate climates more especially, that the air is usually warmer during a fall of snow, than before or after. This increase of temperature probably arises from the extrication of heat, in the sensible form, during the transition of the vapour from a fluid to a solid state. Snow-water has also been said to contain much oxygen, and thus to be particularly favourable to vegetation.

Sleet, is half melted snow; and constitutes the intermediate condition between that of snow, and that of rain, to be next considered.

Of Rain.—When the temperature of the air is above 32°, the freezing point of water; the water separated from the air falls to the earth in the state of rain. Such is a general expression of the fact; but after all the attention that has been bestowed on the phenomena of rain, many difficulties attend the investigation, which have not yet been surmounted.

It cannot be doubted that rain is in some way connected with change of temperature; the perplexity attending the subject, arises, partly from the impossibility, in many instances, of accounting for the supposed change of temperature ; but much more from the difficulty of understanding how this change of temperature operates. According to the usual opinion, the precipitation of water from the atmosphere, is the effect of the mingling together of currents of warm and of cold air, which are supposed to operate on each other in the following manner :

From the law of the tension of vapour, already described, it follows, that when two currents of air having different temperatures, but both alike saturated with vapour, are mixed together ; though the resulting temperature of the mixture will be the mean of the two, the resulting tension of the vapour will not be likewise the mean. The resulting tension of the vapour will always *exceed* the tension belonging to the resulting mean temperature ; consequently, there will be an excess of vapour, which will be precipitated in the form of water. Thus let us suppose two currents of air, both saturated with vapour, the one having a temperature of 40°, and the other a temperature of 60°; and that these two currents of air are mingled together ;

Inches of Mercury.

The tension or elastic force of vapour at 40° is equal to ·263

.   .   .   .   . of vapour at 60° is . . ·524

·787

Mean Tension . . ·398

Whence it appears, that the mean temperature of the two volumes of air is 50°, and the mean of the elasticities of their vapour ·393 inches. But the actual tension or elastic force of vapour at 50°, is not ·393 inches, but only ·375 inches: after the intermixture, therefore, of the two currents, a quantity of vapour will remain, proportionate to the tension of ·018 inches; and as this superfluity of vapour cannot be held in solution by air of the mean temperature of 50°, it will be separated in the form of clouds, or of rain, according to circumstances.

Such, in few words, are the opinions respecting rain first advanced by Dr. Hutton; and notwithstanding some difficulties about these opinions, there can be little doubt of their general accuracy. The subject of condensation may perhaps receive some additional elucidation, from the principles regulating a mixed atmosphere of vapour and air, which were formerly explained; and which may be thus applied. When two currents of atmospheric air of different temperatures, and each charged with vapour up to the point of saturation, are brought into contact; they begin to intermingle, by virtue of the diffusive tendencies of the air and vapour, and the immediate result will be the formation of *visible vapour*; that is to say, of a *cloud*. If the currents are continuous and uniform, the clouds soon spread in all directions, so as to occupy the whole horizon ; while the additional moisture incessantly brought by the warmer current, keeps up a constant supply for condensation, and produces a great and continued deposition of moisture in the form of rain. By degrees, the currents completely intermingle, and acquire a uniform temperature; condensation then ceases; the clouds are redissolved; and the whole face of nature, after being cooled and refreshed by the necessary rain, is again enlivened by the sunshine, thus rendered still more agreeable, by its contrast with the previous gloom.

In this manner, the principles formerly detailed, may be applied to the explanation of the phenomena of rain ; and as far as the explanation goes, it is perhaps quite satisfactory. It must, however, be allowed, as we have before stated, that the utmost information, which we can at present bring to bear upon the subject of the general condensation of moisture from the atmosphere, and of rain in particular, leaves it involved in considerable obscurity.

The following additional particulars, regarding the *effects of different localities; and of different circumstances in the same locality,* which appear to influence the fall of rain, may interest the general reader.

It has been remarked, that in the greater number of instances, more rain falls in the neighbourhood of the *sea,* than at sea ; a fact easily

understood from the principles which have been
stated. Among *mountains* also, more rain falls
than on plains; the excess is indeed striking.
Thus in our own country, at Kendal and at
Keswick, both inclosed by mountains, the annual
fall of rain amounts to 67½ and 54 inches respectively; while in many inland places, the
quantity of rain that falls in the course of a
year, hardly exceeds 25 inches. So at Paris,
the annual fall of rain is only about 20 inches,
but at Geneva 42½ inches; and on the Great St.
Bernard, the highest meteorological station in
Europe, upwards of 63 inches of rain fall during
the twelve months.

Although more rain falls in mountainous districts than in plains; it has been completely
established, that more rain falls at the foot of a
mountain, than on its top. In general, too, a
larger proportion of rain is separated from the
air, near the *earth's surface*, than at any height
above it; a discrepancy of which the present
extent of our knowledge does not enable us to
give a satisfactory explanation.

In most *Tropical countries*, rain falls only at
particular seasons of the year, there being
scarcely any rain during the other seasons.
Thus at Bombay, the rainy months are June,
July, August, September, and October, while
the other months are almost without rain; but
on the opposite side of India, along the Coro-

mandel coast, the time of the occurrence of the
rainy season is reversed : facts strikingly illustrative of the effect of the intervention of the
high table land that separates the two coasts;
and which probably, by influencing the atmospheric currents, gives rise to this singular alternation of weather.

In *temperate climates*, though the total quantity
of rain that falls be much less than within the
Tropics, there is no protracted dry season ; and
the rainy days in the year are more numerous,
the nearer we go to the Poles. Still in general,
more rain seems to fall in temperate climates
during the last six, than during the first six
months of the year.

Among the circumstances which influence the
quantity of rain in the same locality, the most
remarkable are *diminution of temperature*, and
the *unusual prevalence of certain winds.*

With respect to *diminution of temperature*, it
has been observed that almost all wet seasons,
or at least wet summers, in temperate climates,
are unusually *cold*. Now, from the principles
formerly advanced it will be easily understood,
how a depression of the temperature below the
general standard in any locality, may give rise
to a greater precipitation of moisture in that
locality. The locality which has become colder
than those around it, acts as a refrigeratory, and
not only condenses and thus deprives of their

elastic force, all the vapours which are in contact with it ; but the neighbouring vapours rush towards the colder locality as towards a vacuum, either in the form of visible vapour or clouds, in which case they are carried by the winds; or as invisible vapour, in which form their movement may be determined by diffusion.

The effect of the *unusual prevalence* of certain winds in producing an increase of rain, or the reverse, is well known, and is quite intelligible on the principles we have explained. Thus in tropical climates, during the steady prevalence of the trade winds, the currents intermingle but little, the atmosphere is perfectly cloudless, and no condensation takes place. But when these great currents, following the course of the sun, begin at certain seasons of the year to shift their direction ; their uniform course suffers derangement, they become intermixed, and condensations of moisture commensurate with the high temperature, are produced to an extent quite unknown in temperate climates. These condensations form the violent periodical rains of hot climates. So also in temperate climates, as for instance in our own country, winds coming from the south and from the west are from a warmer climate, and hold much vapour in solution; while winds from the opposite points are colder, and are therefore relatively drier. Hence winds from the south and from the west, are more frequently accompanied by rain, than winds from the north and from the east : though as we might expect, the precipitation of rain is most decided, during the conflicts between these opposite currents, which sometimes extend over a large tract of country. The long prevalence of certain winds may thus cause the seasons to be wet in one part of the world, and dry in another; the water being as it were, distilled off from the one, in order that it may be precipitated on the other. Yet the whole amount of the rain in the two countries may perhaps differ very little from the usual average, while the two countries have the benefit of variety in the general amount of their rain ; which variety may be salutary at particular periods, and may even be necessary to their well being.

Before we end the examination of the phenomena of rain, it may be proper to advert to the generally admitted influence of the *Moon* on the weather, and especially on the fall of rain. This influence, however, can hardly, in the present state of our knowledge, be brought to elucidate the phenomena of rain ; so great are the disturbing effects of local and other peculiarities.

Of Hail.—The last form in which we have to consider the precipitation of water from the atmosphere, is hail. Hail may be regarded as consisting of drops of rain, more or less suddenly frozen by exposure to a temperature below 32°.

If the degree of cold has been very sudden and intense, which is often the case, the icy nucleus, from its being of a temperature far below the freezing point, acquires magnitude as it descends, by condensing on its surface the vapour of the lower regions of the atmosphere. Hence, even under ordinary circumstances, hailstones often become of considerable size, are nearly always more or less rounded, and when broken, are seen to be composed of concentric layers.

From what has been stated it will be readily inferred, that hail is not a product of extreme climates; indeed hail may be said to be peculiar to temperate climates, as it rarely ever occurs beyond the latitude of 60°. Hail is most frequent in spring and in summer, when it is often accompanied by thunder. It seldom hails in winter; and hail during the night is very uncommon. In tropical countries there is little hail in any place that is not more than 2000 feet above the level of the sea: in temperate climates, on the contrary, mountain tops are almost free from hail. Certain countries, especially some parts of France, are very liable to hail storms; and such is at times the fury of these storms that they lay waste whole districts. There are on record many instances of these calamitous visitations; which are usually accompanied by whirlwinds, and by the most appalling electrical phenomena.   During storms of such degree of severity, hail-stones have sometimes fallen of enormous magnitude, and often of an irregular shape, as if they were the fragments of a thick sheet of ice suddenly broken : a supposition which alone will explain the formation of angular masses, many inches in size, and many pounds in weight.   The production in the middle of summer of the intense cold which is thus indicated, is a puzzle, philosophers have been unable to solve.

Of the actual Quantity of Water that is evaporated and condensed over the Globe.— Before we close the subjects of evaporation and condensation, it remains to make a few observations on the actual quantity of water that is evaporated and condensed over the globe.

From the principles we have stated, it will appear, that the quantity of water evaporated and condensed over the globe, may be supposed to vary with the mean temperature, and consequently with the latitude.   The following table shows the general truth of this supposition ; and that the average quantity of rain diminishes from the Equator to the Poles.   In fact, a much larger quantity of rain must fall in the Equatorial, than in the Polar regions, as is sufficiently proved by the magnitude of the rivers within the Tropics ; for the size of the rivers of course depends on the quantity of the rain ; the rivers being the conduits along which a certain portion of the precipitated water is borne to the sea.

TABLE.

	Inches.
Uleaborg	13·5
Petersburg	16, 17·5
Paris	19·9
London	*20·7, †22·2, ‡25·2
Edinburgh	22·, 24·5, §26·4
Mean of Carlsruhe, Manheim, Stuttgard, Wurtzburg, Augsburg, and Regensburg, (Schow)	25·1
Epping	27·0
Bristol	29·2
England (Dalton's mean)	31·3
Liverpool	34·1
Manchester	36·1
Rome	39·0
Lancaster	39·7
Geneva	42·6
Penzance	44·7
Kendal	53·9
Mean of twenty places in the lower valleys at the base of the Alps	58·5
Great St. Bernard	63·1
Vera Cruz	63·8
Keswick	67·5
Calcutta	81·0
Bombay	82·0
Ceylon	84·3
Adam's Peak, ditto	100·
Coast of Malabar	123·5
Leogane, St. Domingo	150·‖

---

* Dalton.   † Daniell.   ‡ Howard.   § Adie.

‖ From the Encyclopædia Metropolitana. Article Meteorology, p. 123.

In this table, the names of the places to which
it refers are arranged progressively, according to
the amount of rain that falls in each place; and
though the progression exhibits great irregularities, yet the table fully establishes the general
decrease of rain with the increase of distance
from the equator.

Sir John Leslie has shown, that if all the
aqueous vapour which can at any time be held
in solution by the whole atmosphere, were at
once precipitated on the earth in the form of
rain, it would not be more than about five inches
in depth : now, as in the course of a year, many
times this quantity of rain fall from the atmosphere, its replenishment, of course, must depend
upon evaporation ; of which evaporation we may
thus infer the general amount. With respect
to the quantity of rain which descends annually
on the entire surface of the earth, we want the
means of forming an estimate ; though there is
no proof that this quantity is subject to any
material difference. The distribution, indeed,
as we have seen, diminishes with the latitude,
and varies according to numerous local peculiarities ; some of which have been pointed out in
the preceding paragraphs. Often also, no doubt
for the wisest purposes, the same place is liable to
great fluctuations in the annual amount of rain,
or at least in the times of its precipitation. Yet
all these variations oscillate within certain limits,

and scarcely affect the mean quantity proper to the place; thus showing, that the distribution of rain obeys the same laws which regulate the more fixed operations of nature.

Of the whole water that is condensed upon the surface of the earth, a certain portion, of course, enters into the soil. The depth to which such water sinks, is determined by the declivity of the surface, by the nature of the inferior strata, and by other circumstances; but, usually, after a greater or less period, and range of circulation, the water again makes its appearance in the open day, in the form of *Springs*. The conjunction of springs and the occasional addition of a portion of rain water, which is neither immediately. absorbed by the soil, nor evaporated, constitute brooks and rivulets; these again uniting, in their progress from the higher and interior parts of the countries where this water has been deposited, form the larger rivers; which, after dispensing innumerable benefits to the inhabitants of the plains in their course, finally discharge their superfluous waters into the ocean. As the origin of the superfluous water which flows from the rivers to the ocean, is thus, unquestionably, derived from the vapour condensed in the interior of the countries where the rivers originate;· it follows, that in every country where there are rivers, *condensation must surpass evaporation*. That is to say, a large proportion of water condensed on the land, must have been evaporated not from the land, but from the neighbouring ocean.

The relative proportions of the water that is condensed, and of the water that is evaporated, vary exceedingly in different countries. Such indeed is the amount and variety of the differences, that it is impossible to estimate them; though it is probable that in the same country, the proportions are nearly constant; or, at least, that there is a mean proportion, about which the differences oscillate within trifling limits. In this country, Dr. Thomson has estimated, that taking the whole of Great Britain together, the mean fall of rain amounts, in the course of a year, to 36 inches, the dew being included, (which is considered to amount to about four inches); and that the quantity of water evaporated is about 32 inches. Consequently, the excess of four inches must be supposed to go to supply the springs and rivers; and as these four inches are thus not taken up again by evaporation from the land, they must be drawn from the seas that encircle our shores.* These estimates

---

* On heat and electricity, p. 266. It is proper to observe, that this estimate differs considerably from a previous estimate of Dr. Dalton, who fixes the proportion of water as flowing off by the rivers, in England and Wales, at thirteen inches. It is probable that the truth lies somewhere between the two estimates.

P.   z

of the water that is condensed and evaporated in Great Britain, can only be viewed as rude approximations; and, even admitting them to be correct, they could scarcely be applied with any advantage, to an inquiry into the actual condensation and evaporation in other countries or climates; which in all instances, must be determined by observation and experiment.

Before we quit this subject, perhaps it may not be amiss to endeavour to convey to the general reader, some still more definite notion of the enormous quantity of water which falls from the atmosphere to the earth. Let us suppose an area of nine square miles, which is considerably less than that occupied by London; and that in the course of the year, all the rain which falls in that area, if it stagnated and no evaporation took place, would cover the earth to the depth of two feet; which is about the quantity, as we have seen, that annually falls in London. According to these suppositions, there must fall in London no less than 59,584,084 hogsheads of water in the course of the year; or 163,244 hogsheads daily; the whole of which in the limited space of nine square miles, must have been dissolved in the atmosphere, or suspended in the form of clouds.

Fourthly, *Of the Distribution of Heat and Light in their latent and decomposed Forms through the Vapour of the Atmosphere; and of the Effects of that Distribution.*—The general distribution of heat and light in their latent form through the vapour of the atmosphere, seems to follow the same laws as the distribution of sensible heat formerly explained. That is to say, the distribution of these forms of heat and light decreases from the Equator toward the Poles. The most remarkable effects of the distribution of latent heat have already been incidentally mentioned, and need not be here repeated. We shall therefore proceed to consider the particular distribution of electricity, and of the decomposed forms of light, in the vapour of the atmosphere; and the effects of this distribution.

Of the Relations of Electricity to the Vapour of the Atmosphere.—Atmospheric air, when perfectly dry and pure, is one of the most complete non-conductors of electricity that are known. Whether water in the state of vapour possesses similar non-conducting properties does not appear to be satisfactorily established. But the non-conducting powers of aqueous vapour must be very considerable; otherwise, since the atmosphere is never entirely free from vapour, electrical insulation could not take place. On the other hand, when vapour assumes the form of water, the water instantaneously becomes a conductor of electricity. Hence a mass of visible vapour or cloud, when floating in a mixed atmosphere of air and vapour, is perfectly insulated, and is thus capable of electrical accumulation. Now the phenomena arising from the equalization of such derangements of electrical distribution, are *lightning* and *thunder*. Lightning and thunder therefore are nothing, either more or less, than the phenomena of electricity on a large scale; that is to say, a cloud and the earth, or two clouds, become surcharged with the two opposite forms of electricity, and thus represent the interior and the exterior coatings of an electrical jar similarly surcharged: the intervening and non-conducting air are represented by the interposed and non-conducting glass; while the lightning and the thunder are the spark and the explosion caused by the union of the two electricities. If the reader bear in mind this analogy, it will enable him to understand the whole electrical phenomena of the atmosphere.

The distribution of electricity, like that of heat and light, decreases from the Equator toward the Poles. Thus, in intertropical countries alone, are the effects of this energetic agent displayed in all their power; there, thunder storms are quite terrific, and far surpass any thing of which those, who have not witnessed them, can form a conception. In temperate climates the effects of atmospheric electricity are usually most severe in the summer; and their severity is greater in mountainous districts than in plains. Yet, even under these circumstances, they are much subdued, as compared with what takes place between the Tropics; while in the Polar regions electrical phenomena are still less striking.

Notwithstanding, however, that the general distribution of electricity in the atmosphere, evidently follows the general distribution of sensible heat, it is a remarkable fact, that whenever electrical phenomena are more than ordinarily vehement, they are accompanied by some unusual appearance of *cold*. Thus the alarming descents of hail formerly noticed, which occur most generally in temperate climates, have, in nearly every instance, been attendants of violent thunder storms. Snow also is almost always highly electric. These, and many other circumstances connected with the great and sudden production of cold in the higher regions of the atmosphere, during the display of electrical agency, cannot, in the present state of our knowledge, be explained. For example, whence, in the middle of summer, arises that instantaneous developement of extreme cold, which occasionally produces the terrific hailstorms above alluded to? At present the answer does not appear. Whether the principles advanced in the present volume be capable of solving the difficulty, time must determine.

With respect to the *sources* of the electricity of the atmosphere there have been many opinions. It seems now to be admitted that electrical excitement does not arise from the mere evaporation and condensation of water; but that in order to produce such excitement, there must always be some chemical combination or separation.* Thus electrical excitement is the result of the chemical changes which often accompany the evaporation of water. During combustion also, there is an ample evolution of electricity; the burning body giving out negative, the oxygen positive electricity. In like manner, the carbonic acid sent forth during vegetation is charged with negative electricity; and at the same time the oxygen, as is most likely, is charged with positive electricity. Derivation from these sources has been deemed quite sufficient to explain the very large quantities of electricity, which are so often accumulated in the clouds. It is however probable that there are yet other causes, or at least *one* other cause, on which, in numerous instances, this accumulation may still more immediately depend. We allude to a supposed combination of oxygen with the vapour of the atmosphere. For reasons which we cannot here detail, our opinion is, that this supposed combination of aqueous vapour with oxygen, more

* Pouillet, Elémens de Physique expérimentale et de Météorologie, tom. ii. p. 823.

than any other cause whatever, is in some way concerned with the phenomena of atmospheric electricity. See Appendix.

The *Aurora borealis* is a phenomenon supposed to have some connection with electricity; though its precise nature is involved in considerable obscurity. The phenomenon evidently indicates currents of some kind; and if the light be electrical, we can only suppose such electrical currents to take place in an imperfectly conducting medium. That is to say; if the phenomenon, as some contend, exist in the lower regions of the atmosphere.; luminous electrical currents can be produced only by water in the liquid state: if the phenomenon exist in the higher regions of the atmosphere, as at present is supposed; such currents may depend upon the extreme tenuity of the atmosphere in these higher regions. Our own opinion is, that at different times, the *aurora borealis* exists at different heights in the atmosphere, and consequently may depend upon both these causes.

The phenomena depending upon the decomposition, refraction, and reflection of light by the vapour of the atmosphere are not less striking and important, than those produced by electricity. To such effects upon light by the atmospheric vapour, we owe not only the cærulean tint of the sky, and all the splendid colouring of the clouds; but the beneficial morning and evening

twilight; nay even the light of day itself. "Were
it not," says Sir J. Herschel, "for the reflecting and scattering power of the atmosphere, no
objects would be visible to us out of direct sunshine, every shadow of a passing cloud would
be pitchy darkness; the stars would be visible
all day, and every apartment into which the sun
had not direct admission would be involved in
nocturnal obscurity." Again, to use the words
of the same author, in speaking of twilight,—
" After the sun and moon are set, the influence
of the atmosphere still continues to send us a
portion of their light ; not indeed by direct transmission, but by reflection upon the vapours and
minute solid particles which float in it, and perhaps the actual material atoms of the air itself."*
Such are the beautiful phenomena, and the important results, of the action of the vapour of the
atmosphere on light. It remains to mention a few
others, of a similar character, and produced by
the same causes; but of less frequent occurrence,
or of less importance in the economy of nature.

The first of these minor phenomena which we
shall notice is the *Mirage;* a phenomenon depending partly on the vapour of the atmosphere, and partly on the intermixture of strata
of air of different temperatures and densities.
The mirage is not unfrequent in level countries,

* Treatise on Astronomy, p. 33.

when their surface is strongly heated by the sun's rays, and evaporation results from the continuance of the heat. The mirage assumes the appearance of a sheet of water, often exhibiting the reflected or inverted images of distant objects. In Egypt and in the neighbouring sandy plains, where the mirage is very common, the illusion is at times so perfect, that travellers can hardly be convinced of the non-existence of what they imagine they see.* The phenomena are quite explicable on well known optical principles.†

Nearly allied to the mirage, is the appearance termed *Fata Morgana,* which is occasionally witnessed in the Straits of Messina. There are many similar phenomena, all of them owing to the refraction of light by media of various densities.

The next class of phenomena to be noticed, are those produced upon light by crystals of ice floating in the atmosphere; or by visible vapour. The angular forms of the crystals of ice, by determining the rays of light in different directions, give origin to various eccentric *halos;* which, by their united intensities, particularly where they cross one another, occasionally produce conspicuous masses of light, denomi-

* See Clarke's Travels.
† See Wollaston, Philosophical Transactions, 1800, p. 239.

nated *parhelia* and *paraselenes,* or *mock suns* and *mock moons.* Visible vapours, consisting of water in the fluid state, likewise sometimes form *halos;* but these halos (when more than one exists) are always concentric, the sun or moon being in the centre. These two phenomena not unfrequently take place at the same time.

The last and most frequent phenomenon of the general kind which we shall notice, is produced by the action of *fluid drops* of water upon light, viz. the well known phenomenon termed the *Rainbow.* The concomitants of the rainbow are familiar to every one : there must be rain along with sunshine. Under these circumstances, if the spectator turns his back to the sun, he sees the coloured bow projected on the opposite cloud, and displaying all the tints of the prismatic spectrum.

We are informed in the sacred narrative, that this beautiful phenomenon was chosen as a symbol to mankind of their exemption from future deluge. The sceptic may be challenged to state what pledge could have been more felicitous or more satisfactory. In order that the rainbow may appear, the clouds must be *partial.* Hence the existence of the rainbow is absolutely incompatible with *universal* deluge from above. So long, therefore, as " He doth set his bow in the clouds ;" so long have we full assurance that

these clouds must continue to shower down good
and not evil upon the earth.

3. *Of the Occasional Presence of Foreign
Bodies in the Atmosphere; and of their Effects.*—
The foreign bodies which occasionally exist in
the atmosphere, may be considered as of two
kinds, viz. *those which are merely suspended in
the atmosphere in a state of mixture;* and *those
which pervade the atmosphere in a state of
solution.*

Both in ancient and in modern times, and
in various parts of the world, rain and snow
have been observed to be coloured by an admixture of extraneous matters.   The nature
of these colouring matters has been found to
be very different in different instances.   Some
have proved to be of vegetable origin, consisting
of *minute lichens* and *other cryptogamous plants*
brought from a distance by the agency of the
winds, and diffused in myriads through the
atmosphere.   Such vegetable matters have been
sometimes more or less *red:* whence those imagined *showers of blood* we read of as producing
so much popular excitation.   In other instances
the colour has been given by earthy and
metallic matters in a state of very fine powder;
and in this case their descent has been usually
accompanied by violent electrical phenomena,

similar to the phenomena which almost always attend the descent of *Meteoric stones* or *Aerolites*, to which perhaps they are nearly allied.

Of the falling of stones from the atmosphere, there can now be no doubt; though the origin and the nature of these stones are very obscure, and indeed cannot, in the present state of our knowledge, be explained. There have been various opinions on the subject. Some, considering aerolites to be the productions of our own planet, have viewed them as masses projected from volcanoes to a great height and distance in the atmosphere; or as formed by the agglutination of the earthy and metallic powders from volcanoes, as before mentioned. Others ascribing to aerolites quite a different origin, have viewed them as fragments scattered through space, which happening to come within the sphere of our earth's attraction, have been determined to its surface, &c.

Although we are thus uncertain regarding the origin of aerolites, or their use in the economy of nature; it now seems by innumerable observations to be completely established, that aerolites, while in the higher regions of the atmosphere, are often in a state of intense ignition. They there assume the form of brilliant meteors, which as they approach the earth, burst with a loud explosion, followed by a shower of stones. These stones generally exhibit evident marks of fusion; and many of them have been picked up while still warm, so as to leave no doubt of their being real aerolites. It is singular too, that the composition of aerolites is in some degree peculiar. They invariably contain, either iron, or cobalt, or nickel, or all these three metals, in union with various earthy substances. Aerolites have been found of every size, from that of a few grains to the weight of several hundreds of pounds; for of this weight are some of those isolated masses of iron seen in different parts of the world, and which are generally allowed to be of meteoric origin.

Intermediate, as it were, between substances suspended, and substances dissolved, in the atmosphere, are those matters, whatever their nature may be, which have been known to spread as a haze over large districts, and have been termed " *Dry Fogs.*"

In the year 1782, and still more in the year following, a remarkable haze of this kind extended over the whole of Europe. Seen in mass this haze was of a pale blue colour. It was thickest at noon, when the sun appeared through it of a red colour. Rain did not in the least degree affect it. This haze is said to have possessed drying properties, and to have occasionally yielded a strong and peculiar odour. It is also said to have deposited in some places a viscid liquid, of an acrid taste, and of an unpleasant smell. About the same time, there were, in Calabria and in Iceland, terrible earthquakes, accompanied by volcanic eruptions. These earthquakes and eruptions were supposed to have been connected with the haze. Indeed it has been generally remarked, that such a condition of the atmosphere has been usually preceded by an earthquake, either in the same or in some adjoining country. The dispersion of this haze in the summer of 1783 was attended by severe thunder storms. As might be expected, the general state of health has, for the most part, been deranged during the continuance of these phenomena. Simultaneously there have been epidemic diseases of various kinds. Thus, in the above mentioned years, 1782 and 1783, an epidemic catarrh, or influenza, prevailed throughout Europe; affecting not only mankind but likewise other animals.*

The nature of the matter thus diffused through the atmosphere is quite unknown. It may, at different times, be not less various than the character of the epidemics to which it gives origin. As an example of the extraordinary effects which foreign bodies, when diffused through the atmosphere, are capable of producing, we may mention those produced by Selenium when, in combination with hydrogen, it is diffused as a gas through

* See article Influenza, in the Encyclopædia of Practical Medicine.

the air, even in the most minute quantity. The effects of this gaseous combination of selenium with hydrogen are thus described by the celebrated chemist, Berzelius, its discoverer. " In the first experiment which I made on the inhalation of this gas, I conceive that I let up into my nostrils a bubble of gas, about the size of a small pea. It deprived me so completely of the sense of smell, that I could apply a bottle of concentrated ammonia to my nose without perceiving any odour. After five or six hours, I began to recover the sense of smell; but a severe catarrh remained for about fifteen days. On another occasion, while preparing this gas, I became sensible of a slight hepatic odour, because the vessel was not quite close; but the aperture was very small, and when I covered it with a drop of water small bubbles were seen to issue, about the size of a pin's head. To avoid being incommoded with the gas, I put the apparatus under the chimney of the laboratory. I felt at first a sharp sensation in my nose; my eyes then became red, and other symptoms of catarrh began to appear, but only to a trifling extent. In half an hour, I was seized with a dry and painful cough, which continued for a long time, and which was at last accompanied by an expectoration, having a taste entirely like that of the vapour from a boiling solution of corrosive sublimate. These symptoms were removed by the application of a blister to my chest. *The quantity* of Seleniuretted Hydrogen Gas, which on each of these occasions entered into my organs of respiration, *was much smaller* than would have been required of any other inorganic substance whatever, to produce similar effects." *

As we have already stated, selenium is for the most part found in association with mineral sulfur. Selenium is also, like sulfur, a volcanic product. Now, though we can hardly imagine the possibility of the diffusion of selenium through the atmosphere in combination with hydrogen ; selenium may be so diffused, in some other form of combination, which may produce effects analogous to those of seleniuretted hydrogen. We do not mean to assert that the diffusion of any such substance really takes place. Our intention is merely to show, that a small quantity of an active ingredient, like selenium, is sufficient to contaminate the atmosphere over a wide extent of country. Such a substance being ejected from the crater of a volcano during an eruption, or through a crevice in the earth during an earthquake, may thus produce an epidemic disease. Nor is it improbable that many epidemics, particularly those of a catarrhal kind, have so originated.

---

* Annals of Philosophy, Old Series, vol. xiv. p. 101.

The matters occasionally diffused through the atmosphere, which appear to be *in a state of solution*, are not often perceptible by our senses, unless in some cases, perhaps, by the sense of smell.

As an instance of the presence of such bodies in the atmosphere, we may mention a very remarkable observation which occurred to the writer of this treatise, during the late prevalence of epidemic cholera. He had for some years been occupied in investigations regarding the atmosphere; and for more than six weeks previously to the appearance of cholera in London, had almost every day been engaged in endeavouring to determine, with the utmost possible accuracy, the weight of a given quantity of air, under precisely the same circumstances of temperature and of pressure. On a particular day, the 9th of February, 1832, the weight of the air suddenly appeared to rise above the usual standard. As the rise was at the time supposed to be the result of some accidental error, or of some derangement in the apparatus employed; in order to discover its cause, the succeeding observations were made with the most rigid scrutiny: but no error or derangement whatever could be detected. On the days immediately following, the weight of the air still continued above the standard: though not quite so high as on the 9th of February, when the change was first noticed. The air retained its augmented weight during the whole time these experiments were carried on, namely, about six weeks longer. The increase of the weight of the air observed in these experiments was small; but still decided, and real. The method of conducting the experiments was such as not to allow of an error, at least to an amount so great as the additional weight, without the cause of that error having become apparent. There seems, therefore, to be only one mode of rationally explaining this increased weight of the air at London in February, 1832; which is, by admitting the diffusion of some gaseous body through the lower regions of the atmosphere of this city, considerably heavier than the air it displaced. About the 9th of February, the wind, which had previously been west, veered round to the east, and remained chiefly in that quarter till the end of the month. Now, precisely on the change of the wind, the first cases of epidemic cholera were reported in London; and from that time the disease continued to spread. That the epidemic cholera was the effect of the peculiar condition of the atmosphere, is more perhaps than can be safely maintained; but reasons, which have been advanced elsewhere, lead the writer of this treatise to believe, that the virulent disease, termed cholera, was owing to the same matter which produced the additional weight of

the air. The statement of these reasons here
would be quite out of place : it is enough to say,
that they are principally founded on remarkable
changes in certain secretions of the human body,
which, during the prevalence of the epidemic,
were observed to be almost universal; and that
analogous changes have been observed in the
same secretions of those, who have been much
exposed to what has been termed *Malaria*.
The foreign body, therefore, diffused through the
atmosphere of London, in February, 1832, was
probably a variety of malaria, a subject which we
now proceed to consider.

In districts partially covered with water, and
having a luxuriant vegetation, such as marshes
and fens, particularly in warm countries ; or in
colder countries, at seasons of the year when the
sun is most powerful ; noxious exhalations arise,
whose nature differs perhaps in some degree according to the locality. Such exhalations have
received the general name of Malaria, and are
well known to be the fertile source of various
diseases, more or less, of the intermittent febrile
type. In cold and in temperate climates, these
diseases for the most part assume the character
of regular ague, or of rheumatism : but on
approaching to, and within the Tropics, they
appear as the more formidable remittent and
continued fevers, the well-known scourges of
hot climates.

With respect to the nature of these exhalations, our knowledge is very imperfect. Evidently, they are in some way connected with vegetation; not however with vegetation living and in a state of growth, but with vegetation in a state of decay. It has therefore been thought likely, that these exhalations contain some gaseous body, composed chiefly of hydrogen and carbon. Their effect may arise from a gaseous compound of this description; though no such compound is at present known : and the probability is, that malaria occasionally owes its properties to other elements, besides the hydrogen and carbon disengaged from decaying vegetables.

We have thus endeavoured to give a concise statement of that wonderful assemblage of Laws, of Adaptations, and of Arrangements, which viewed together constitute what we term *Climate;* and which, as affecting the welfare of the denizens of this globe, undoubtedly, are not surpassed in interest or importance by any others throughout the whole of nature. Of the innumerable suns and planets which may occupy the boundless expanse of the universe, we feel not the influence ; even their existence scarcely obtrudes itself upon our notice. But in the light and the heat of our own sun, and in the wind and the rain of our own atmosphere, every organized being on this earth, from Man, the

Lord of its creation, down to the humblest plant that drinks the dew, is alike most intimately concerned. The subject of Meteorology, therefore, in all ages and countries, has attracted the especial attention of mankind. In ruder states of society, empirical prognostics, founded on the aspect of the clouds, on the movements of animals, and on other incidental occurrences, formed the study of those who pretended to a fore-knowledge of the weather ; while electrical phenomena were objects of superstitious awe. In modern times much of this wonder and uncertainty has been removed. The gloom or the clearness of the air ; the mists and the halos of a stormy sky; the restlessness and clamour of animals ; &c., are now referred simply to that overcharge of moisture, and to that unequal distribution of electricity, which precede a fall of rain. Nay, the very thunderbolt has been arrested in its course, and being no longer an object of amazement, has been divested of half its terrors.

But is this advance in knowledge calculated to lessen our veneration for the great Author of Nature, or to derogate from his wisdom and his power? On the contrary, our estimate of both must be greatly increased. Of the Deity, infinite as he is, and dwelling in infinity, we finite beings can form no conception. What little, therefore, we can know of Him, we know nearly altogether from his works. Consequently, whoever has most studied His works, will be the best qualified—nay, will be alone qualified, to form an adequate conception of Him. Thus, to measure, to weigh, to estimate, to deduce, may be considered as the noblest privileges enjoyed by man ; for only by these operations, is he enabled to follow the footsteps of his Maker, and to trace His great designs. Instructed by these operations, he sees and appreciates the wisdom and the power, the justice and the benevolence, that reign throughout creation : he no longer gazes on the sky with stupid wonder ; nor dreads the thunderbolt, as manifesting the wrath of a vengeful Deity.

The constituents of climate, even imperfectly as they can be understood by us, are seen to be adjusted and arranged in a manner so surprising ; that to those who admit the existence of design, they require only to be stated and apprehended, in order to their being received as additional proofs of that great argument. Where all are great, and splendid, and good, selection is precluded : but the circumstances accompanying the distribution of water over this globe, more perhaps than any other, arrest our notice as indicative of design. Leaving out of view the other properties of water ; on what other supposition, besides that of design, can we account for all these astonishing properties, on which depend its evaporation and diffusion through the atmosphere—its subsequent condensation, not at once in the form of water or of ice, but in the intermediate state of clouds—its colour and lightness when in the state of snow—its power of refracting light and of conducting electricity —in short, all the numerous, minute, and happily contrived qualities displayed by this highly elaborated fluid ? These qualities together, form such a union of adaptations and arrangements, each most successively accomplishing a particular purpose, and apparently directed to, and designed for, that purpose ; that to doubt the agency of design would seem impossible. Yet some men's minds are so warped, that they either cannot, or will not, be persuaded of the existence of design ; but asserting the omnipotence of the laws of nature, they forget Him who framed these laws, and are reluctant to give credence to His being, or to His power. To such persons, Meteorology offers one or two exclusive arguments, which, at the risk of being accused of tediousness, and unnecessary repetition, we shall urge briefly in this place.

The great Author of Nature, as we have before said, has chosen to act agreeably to certain established laws, by which he is invariably guided. Some of these laws we are able more or less to comprehend; and we can refer them to more general principles. Others are beyond our comprehension : we see only their effects ; and even these effects are most imperfectly revealed to us. As instances of the laws of nature which it is in our power to refer to general principles, may be mentioned the currents in the ocean and in the atmosphere, by which the equilibrium of temperature over the globe is maintained. These currents, we know, are strictly referrible to hydrostatic and pneumatic principles. The argument of design, which is deducible from these principles, rests, therefore, not so much on the principles themselves, as on their application precisely where they are requisite. On the other hand, as we stated at the commencement of this treatise, the laws of chemistry are founded solely on experience ; so that our acquaintance with them is very defective : for in very few instances are they referrible to the laws of quantity; and even when they can be so referred, it is only in a manner very imperfect. But though we do not comprehend the laws of chemistry, we see that many of them, perhaps all, in so far as they are intelligible to us, are entirely consistent with each other ; and are as uniform in their operation as those which obviously depend on mechanical principles, or on the laws of gravity. Thus the laws, that all bodies are expanded by heat and are contracted by cold—that chemical substances do not mix, but always combine in certain proportions, and in no others,—are general laws, to which there are so few exceptions, that they are calculated on almost with as much certainty, in the operations of nature, and in the common intercourse of mankind; as the invariable and necessary results, that a heavy body will fall to the earth, or that two and two make four. We have selected these laws of chemistry, partly from their general and indisputable character, and partly that the force of the argument which follows may be more conspicuous.

All bodies are expanded by heat and are contracted by cold. If water had not constituted an exception to this law, though all its other properties had been the same as they now are, long before this time, as we have seen, half the water on the globe would have been converted into ice; and the existence of organized beings would have been physically impossible.

All chemical substances combine in certain proportions, and in no others. If air had been formed according to this law, every thing else being the same as at present, long before this time, half of the air in the atmosphere would have been contaminated, and rendered unfit for the support of animal life. In order, therefore, that *water might not be frozen;* and that *air might not become irrespirable; laws must be infringed*—and THEY ARE INFRINGED; infringed too, precisely where their infringement, both in kind and degree, is indispensably necessary to organic existence. Now, we appeal to the most inflexible sceptic regarding the argument of design and ask him, on what other principle, unless that of express adaptation and design, can two such general laws have been infringed, exactly in those instances in which their infringement is wanted, and no where else? Of the sophistry by which the evasion of this plain question may be attempted, we are quite ignorant. But we cannot resist the conviction, that one purpose of the arrangement has been to confound the presumptuous sceptic; who is thus perpetually reminded of the infringement of his boasted " laws of nature," by the very water he drinks, and by the very air he breathes.

With respect to the foreign bodies in the atmosphere, which have been treated of in the last section, it remains to observe, that though of very opposite characters, they have yet this resemblance; that they all apparently exist less on their own account, than as being the inevitable results of general laws established for a higher purpose. Such results of general laws, may be considered as analogous to the coldness and darkness, which necessarily prevail around the poles, from the earth's position in relation to the sun: and they have been alike permitted; not because they could not have been avoided or removed; but in the language of Paley, before quoted, " because the Deity has been pleased to prescribe limits to his own power, and to work his ends within these limits."

Man, forgetting how insignificant he is, and how limited his utmost knowledge, is too apt to measure Omnipotence by the standard of his own narrow intellect; and to be guided by his own selfish feelings, in judging of the extent of Divine benevolence. That this earth, a minute fraction, as it is, of a great and wonderful system, should be amenable to the general laws by which the whole system is governed, is, at the least, exceedingly probable. Of such general laws, of their changes, of their aberrations, or of their influences, we, situated in this extremity of the universe, cannot see the object. What, therefore, appears to us anomalous or defective, may in reality be parts of some great cycle or series, too vast to be comprehended by the human mind, and known only to beings of a higher order, or to the Creator himself. So again, amidst the desolation of the hurricane, or of the thunderstorm; in the settled affliction of malaria, and in the march of the pestilence; the goodness of the Deity is impugned, his power even, is regarded doubtfully. But what, in truth, are all these visitations but so many

examples of the "unsearchable ways" of the
Almighty ; " He sits on the whirlwind, and
directs the storm :" a hamlet is laid waste ;
a few individuals may perish ; but the general
result is good : the atmosphere is purified ; and
pestilence with all its train of evils disappear.·
Nay, however inscrutable the object of the
deadly malaria itself, do we not see one end
which it serves, namely, to stimulate the reasoning powers, and the industry of man? By his
reason, man has been guided to an antidote
beneficently adapted for his use, which has stript
malaria of half its terrors. By his industry,
the marsh has been converted into fertile land,
and disease has given place to salubrity.

When, therefore, we duly consider all these
things ; when we reflect also on the number,
the properties, the various conditions of the
matters composing our globe ; the wonder surely
is, not that a few of these matters occasionally
exist as foreign bodies in the atmosphere, but
that others of these matters are not at all times
diffused through it, and in such quantity .as to
be incompatible with organic life. Thus, the
original constitution of the atmosphere, and the
preservation of its purity against all these con-
taminating influences, may be viewed as the
strongest arguments we possess, in demonstration of the benevolence, the wisdom, and the

omnipotence of the Deity : benevolence in
having willed such a positive good ; wisdom
in having contrived it ; and omnipotence in
having created it, and in still upholding its existence.

CHAPTER VI.

OF THE ADAPTATION OF ORGANIZED BEINGS TO
CLIMATE ; COMPREHENDING A GENERAL SKETCH
OF THE DISTRIBUTION OF PLANTS AND ANIMALS
OVER THE EARTH ; AND OF THE PRESENT POSI-
TION AND FUTURE PROSPECTS OF MAN.

IN the general survey of climate, and of its
reference to organization, given in the preceding chapter, we have seen, on the one hand,
that, by a series of wonderful expedients, the
climate or temperature of the greater portion
of the earth's surface, has been so equalized, as
to be brought within the range of organic existence. On the other hand, we shall find, that
by a series of expedients, not less wonderful,
organic existence has been so diversified and
extended, as to include all the possible varieties
of soil and climate. Hence, the arrangement
taken altogether, presents us with such extraordinary instances of mutual adaptation of its va-

rious constituents to each other, as to admit of explanation, only upon the supposition of the whole being different parts of the same magnificent Design ; while the infinite variety, where all might have been otherwise, must be considered as equally indicative of the Benevolence, and the Power of the Designer.

Next to Climate, the circumstance in which organized beings are most immediately concerned, is SOIL; a subject already alluded to, but which it will be necessary to illustrate a little further before we proceed.

The soil is that collection of matters, more or less in a state of comminution, which immediately covering the general surface of the earth, fills up its minor inequalities, and rounds off its asperities. On this layer of comminuted mineral substances and organic remains, all vegetables and animals, at least all land animals, depend for their existence; and, if there ever was a time when the materials composing this globe were collected into solid masses, it is evident that such a condition must have excluded the greater number of plants and animals; even if every thing else had existed the same as at present.

The formation of the soil has apparently been a work of time, and the result of the gradual attrition of the solid materials composing the crust of this globe. Hence the formation of soil has probably been always progressive, and is still going on. Besides this gradual attrition, the harder materials of our globe seem to have suffered much disintegration, during those periodic convulsions formerly mentioned. By the same convulsions, also, the different comminuted materials have evidently been mixed and scattered, and finally deposited over the surface of the whole earth; so as to give occasion to that infinite variety which every where prevails.

The foregoing remarks naturally lead to the conclusion, that the characters of the soil will generally agree with those of the rocks composing the crust of the earth; and this inference is correct. The more common ingredients in rocks are silex, alumina, lime, magnesia, and iron; and these mineral matters actually constitute the greater bulk of every soil. The remaining matters consist of more or less of various other earthy and saline principles, and of vegetable and animal remains.

After these general observations upon soils, we come to the proper subject of this chapter, which we shall consider under the three following sections :—*Of the Distribution of Plants over the earth ;—Of the Distribution of Animals over the earth ;* and,—*Of the present Condition and future Prospects of Man.*

Section I.

Of the Distribution of Plants over the Earth.

FROM what has been stated, it will be readily understood that Soil and Climate are the two great and immediate causes, by which vegetable and animal existence are likely to be affected. We shall, therefore, in the first place, take a view,

1. *Of the Differences of Vegetation, as liable to be influenced by Soil, and by other minor Local Circumstances, in the same Climate.*—The most incurious observer, in travelling through a country, must be struck with the different vegetation that prevails in different parts of the country; and with the effect which this difference produces on the manners and on the health of the inhabitants. Thus, in some parts of England, the *Apple* and the *Pear* are seen growing spontaneously in every hedge-row; while, in other parts, apple and pear trees will not flourish, even with the utmost care. Some situations are favourable to the *Oak*, others to the *Beech*, others to the *Elm*. Accordingly, these well-known and beautiful

trees predominate in some districts, almost to the exclusion of every other, and thus constitute the leading feature in the landscape.

These are familiar examples of partial changes among the larger vegetables of a country ; while the general vegetation is supposed to remain nearly the same. Between such partial change, and the complete establishment of a peculiar vegetation, there exists among different localities, every possible shade of diversity. Many of these differences in vegetation are obviously connected with differences in soil and in situation. Thus, some plants will thrive only on a calcareous soil ; as a few of the *Orchis* tribe in our own country, and the *Teucrium montanum* in Switzerland. Others, like the *Salsolas* and the *Salicornias*, will only grow in salt marshes. Some plants flourish in sea water ; some in fresh : while to others again, water, at least in excess, is so prejudicial, that they can exist nowhere, unless on bare rocks, or in arid deserts. Mountainous situations are most favourable to the increase of some plants ; while others abound in plains. The larger number of plants prefer sunshine, but some are most vigorous in the shade ; and others are so impatient of light, that they are found only where there is absolute darkness. There are, besides, parasitic plants, like the *Mistletoe*, whose nourishment is derived from the plants to which they are attached. In

P.   B B

short, the varieties in the nature of plants are countless, nor is the enumeration of them requisite. What has been stated, is more than enough to show the wonderful arrangements which have been made, to ensure the clothing of every part of the earth's surface with vegetable organization. There is not a soil, however barren, nor a rock, however flinty, that has not its appropriate plant; which plant has no less wonderfully found its way to the spot adapted for it, nay, will perish if removed elsewhere. Saline plants, for instance, will grow only where saline matters are abundant; plants of the marsh, and of the bog, flourish only in marshy and boggy ground; those of the parched desert and of the cloudy mountain, each in its fitting locality. Thus the soil and its occupant seem to have been made for each other; and hence one source of that astonishing variety exhibited in nature.

There are still more remarkable deviations among the plants of different countries remote from one another; even where the circumstances of climate and of soil are in every respect alike. The plants of the Cape of Good Hope, for instance, differ exceedingly from the plants of the south of Europe; though the climate and much of the soil be not dissimilar. Often, on the same continent, nay, on the same ridge of mountains, the plants on the opposite sides have no resemblance. " Thus, in North America, on the east side of the rocky mountains, *Azaleas*, *Rhododen-drons*, *Magnolias*, *Vacciniums*, *Actæas*, and *Oaks*, form the principal features of the landscape; while on the western side of the dividing ridge, these genera almost entirely disappear, and no longer constitute a striking characteristic of the vegetation." *

In general, the plants of America are different from the plants of the old world, except toward the north; where, as might be expected from the near approximation of the two continents, many individuals are common to both. The plants of islands, and those growing in isolated situations are often quite peculiar. Thus the plants of New Holland, with comparatively few exceptions, differ from the plants of all the rest of the world; and, " of sixty-one native species, in the little island of Saint Helena, only two or three are to be found in any other part of the globe." †These facts are quite inexplicable upon any known principles; and are calculated to excite a more than ordinary degree of attention, as being solely referrible to the will of the Great Creator; who has chosen to provide infinite diversity, where all might have been uniform and monotonous; and has thus rendered more

---

* Lindley's Introduction to Botany, page 489.

† See Principles of Geology, vol. ii. by C. Lyall, who has treated this interesting subject in detail. To Mr. Lyall we are indebted for many of the following facts.

conspicuous his wisdom, his power, and his goodness.

2. *Of the Influence of Climate on Vegetation.* —The climate of a place, as has been before shown, independently of minor local causes, is influenced chiefly by the two following circumstances:—the Latitude of the place; in other words, the general portion of heat and light which it receives from the sun;—and its Height above the surface of the sea; by which circumstance of elevation, the heat received from the sun, is liable to be at least as much varied, as by latitude; but the variation is according to other laws than those which depend on mere latitude; indeed, according to laws which vary in different latitudes.

Every one is acquainted with the general fact of the difference between the plants of warm and those of cold countries; between the plants that grow on plains, and those that grow on mountains. Thus, " in the countries lying near the Equator, the vegetation consists of dense forests of leafy *evergreen trees*, *Palms*, and *arborescent Ferns*, among which are intermingled *epiphytal herbs*, and *rigid Grasses*. There are no verdant meadows, such as form the chief beauty of our northern climate; and the lower orders of vegetation, such as *Mosses*, *Fungi*, and *Confervæ* are very rare. As we recede from the Equator, the plants above mentioned gradually give way to *trees* with *deciduous* leaves; rich meadows appear, abounding with *tender herbs;* the *epiphytal Orchideæ* are no longer met with, and are replaced by terrestrial fleshy - rooted species; *Mosses* clothe the trunks of aged trees; decayed vegetables are covered with parasitical *Fungi;* and the waters abound with *Confervæ*. Approaching the Poles, trees wholly disappear; *dicotyledonous* plants of all kinds become comparatively rare; and *Grasses* and *cryptogamic* plants constitute the chief features of the vegetation." *

Changes not very dissimilar are observed in the vegetation at different heights on the mountains of warm climates. Thus, at the base of the celebrated Peak of Teneriffe, the plants have all the distinguishing characters of the plants of Africa. There flourish the .succulent *Euphorbia*, the *Mesembryanthema, Dracœna*, &c.: also the *Date palm*, the *Plantain*, the *Sugarcane*, and the *Indian-fig*. A little higher, grow the *Olive tree*, the *fruit trees* of Europe, the *Vine*, and *Wheat*. Then succeeds the woody region of the mountain; where from the numerous springs the ground is always verdant. In that region is seen a profusion of beautiful evergreens; such as various species of *Laurel*, one of *Oak*, two species of *Iron tree*, an *Arbutus*, and

---

* Lindley's Introduction to Botany, page 484.

several others. Next above, is the region of pines; characterized by a vast forest of trees resembling the *Scottish fir*, intermixed with *Juniper*. Then follows a tract remarkable for the abundance of a species of broom. At last the scenery is terminated by *Scrofularia*, *Viola*, a few *Grasses*, and *cryptogamic* plants.*

The proportions which different groups of plants bear to each other, vary exceedingly in different latitudes. An interesting table given in the Appendix, slightly altered from Humboldt, exhibits the proportional amount of some natural groups of plants, to the whole mass of vegetation in the zones mentioned; and will enable the reader to understand the relation of vegetable forms, to the greater or less distance of their place of growth from the Equator. The arrangement is so obvious as scarcely to require explanation. Thus in the equatorial zone, between 10° north and 10° south latitude, the first group, including *Ferns*, *Lichens*, *Mosses*, and *Fungi*, constitutes on the plains only 1-15th; but on the mountains 1-5th, of the whole number of plants that exist in that zone. While in the temperate zone, the proportion of the first group of plants is at least one-half of the whole number; and in the frigid zone, the entire vegetation consists of plants of that group. The distribution of the other groups is equally remarkable.

* Humboldt.

From this table we learn many interesting particulars, in addition to what has been already stated regarding the distribution of plants over the surface of the globe. We may notice especially the striking difference between the Flora of the Old, and the Flora of the New world, in corresponding parallels of latitude. These differences, in a great many instances, are undoubtedly referrible to the unknown causes to which we have before alluded. But in other instances, they are obviously connected with the difference of temperature prevailing in the two continents under the same parallel. Before we proceed, let us dwell a little longer on the consideration of these beautiful arrangements.

In Tropical countries alone, beneath a vertical sun, do we see vegetation in all its glory and magnitude. There, the form, the colour, and the odour of plants, are developed to the utmost; and where else could they be so developed? where else could the majestic palm rear its towering stem, and send forth its gigantic leaves? where else could we expect to find groves, ever verdant, blooming, and productive? Amidst eternal summer, all this is in character: forests denuded of their leaves, and for half the year assuming the appearance of death, would in such a climate be perfectly incongruous. But in countries remote from the Equator, and in which, during many months, the temperature is more or less depressed, a vegetation thus incessantly active could not exist, nor would it be appropriate. Accordingly, the palm tribe, and many of the more distinguishing productions of the Tropics become gradually fewer in number as we recede from the Equator, and at last give way entirely to deciduous plants; that is, to plants endued with the power of *hybernating*, or sleeping, as it were, in the colder season; and which vegetate only during the warmer portion of the year. And here we have displayed another of those admirable provisions, which at once strike us irresistibly as being the effect of design! In Tropical countries, where the seasons are uniform, and where there is no cold to injure, the leaf buds of plants are without covering or protection, and are freely and confidently exposed to the atmosphere. But in climates where the seasons change, and where vegetation is liable to be suspended by the cold, the leaf buds exhibit a structure remarkably different. Developed in the latter end of summer, or autumn, they are almost invariably provided with *tegmenta*, or coverings; within which, during their period of torpor, they are cradled, safe from cold and from accident!

As we advance still further to the north or to the south, where the winter becomes more severe and of longer continuance, deciduous plants in their turn diminish both in number and in magnitude; and after having shown themselves under a variety of stunted forms, are at length almost entirely superseded by a few coarse grasses and lichens. Yet even here design is apparent. These hardy natives of the poles are, from the simplicity of their structure, wonderfully adapted to the climate of the region they occupy; in which alone they will flourish, and for which alone, therefore, they have been expressly created.

Though it be generally true that plants will grow only in the soil and climate adapted for them; yet, as if intentionally to evince His power, the Great Author of nature has created some manifest exceptions to this rule. All organized beings have been more or less endowed with the faculty of accommodating themselves to circumstances. In the larger number of plants this faculty scarcely exists; but in some it is much stronger; and in others, constituting the exceptions to which we allude, the extent of the accommodating faculty is almost incredible. In general, plants that are the natives of peculiar soils, and of extreme climates, are the most impatient of change; while the natives of ordinary soils, and of temperate climates, have a wider range of growth. The exceptions to the rule of adaptation are chiefly among plants that are natives of such soils and climates. Thus " the *Samolus Valerandi* is found all over the world, from the frozen north to the burning south; associated here with *Birches*, and similar northern forms, and there mixed with *Palms* and the genuine denizens of the tropics. The number of plants, however, which can thus accommodate themselves to all circumstances and climates is limited; while those which readily naturalize themselves in climates similar to their own, are, on the other hand, numerous. Of the latter, indeed, examples present themselves at every step. All the hardy plants, for instance, of our gardens may in some sort be considered of this nature; for although they do not grow spontaneously in the fields, they flourish almost without care in our gardens. The *Pine apple* has gradually extended itself eastward from America, through Africa, into the Indian Archipelago, where it is now as common as if it were a plant indigenous to the soil; and in like manner the *Spices* of the Indies have become naturalized on the coast of Africa and the West India islands. ' To this property of naturalizing themselves, no doubt, is to be referred, in a variety of instances, the presence of the same plants in different countries. For though, as we have just stated, the Flora of different countries is generally different, yet in almost all instances, some plants exist which are found in other countries. Thus, " above 350 species are said to be common to Europe and

North America; and even among the peculiar features of the Flora of New Holland, Mr. Brown recognised 166 European species. The presence of many such strangers may undoubtedly be referred to the agency of man and other animals; to currents in the ocean; to winds; and a variety of natural causes." While " the presence of others, seems inexplicable on any other supposition, than that they have been created in the places where they now exist." *

Hitherto we have considered plants only in relation to the soil, and to the climate, in which they grow; and have not entered into details respecting all the beautiful arrangements, by which their growth has been accomplished. The consideration of these arrangements belongs to the Physiologist, the Botanist, and the Geologist, with whose duties we wish as little as possible, to interfere. There is, however, yet one point of view, in which our argument naturally leads us to consider vegetation; namely, as forming the link, by which animals are connected with the earth; in other words, as furnishing to animals the means of subsistence.

The circumstance, which, perhaps; more than any other, is calculated to arrest our attention with respect to vegetable productions in general, is their vast *profusion*, in every sense of the term;

* Lindley, Introduction to Botany, p. 501.

whether we contemplate their variety, their magnitude, or their number. Thus the numerous and varied plants growing in tropical climates, are equally remarkable from their size, their luxuriant foliage, and the exuberance of their roots and seeds. Let us take, for instance, the palm tribe. It has been estimated that there are a thousand species of palms; and though the number actually known to exist is by no means so large, yet late discoveries seem to render the estimate not improbable. In many of the palm tribe the developement of the form, and the quantity of flowers and fruit is altogether extraordinary. Among others, the species which yields the well known *Cocoa nuts* grows to the height of eighty feet ; each plant flourishes for a century ; and, during the greater part of that time, continues to produce annually at least a hundred of these large nuts. Yet the cocoa nut species may be considered as one of the least productive of the palm tribe : for every bunch of another species, the *Seje* palm of the Oronoko, bears as many as 8000 fruit; while a single spatha of the *Date* palm contains 12,000 flowers ; and in a third species, the *Alfonsia Amygdalina*, there is the enormous number of 207,000 flowers on each spatha ; or 600,000 on a single individual plant!

In superlative exuberance, however, the Palm tribe must yield to the *Banana*, or Plantain, another inhabitant of tropical countries. The fruit of this plant is often a foot in circumference, and seven or eight inches long : it is produced in bunches, containing usually from 160 to 180 fruit; and each bunch weighs from 66 to 88 pounds avoirdupoise. As Humboldt has remarked ; the small space, therefore, of 1000 square feet, on which from thirty to forty Banana plants may grow, will, on a very moderate computation, afford, in the course of a year, 4000 pounds weight of fruit ; a produce 133 times greater, than could be obtained from the same space, if covered with wheat ; and 44 times greater than if occupied by potatoes. The *Orange* tree may be mentioned as another instance of extraordinary fecundity ; thus a single tree at St. Michael's, has been known to bear in a season 20,000 oranges fit for packing, exclusively of those damaged and wasted, amounting to at least one third more. An example to the same effect, but of a different kind, is the *Sugar* cane, which furnishes an unlimited supply of saccharine matter in its purest form ; while various roots, as those of the *Cycas Jatropa,* and many others, abound equally in farinaceous matters.

As we withdraw from the Equator into the regions of hybernating plants, vegetation is seen on a much less magnificent scale ; though in the temperate climates, and even where we might expect to find utter sterility, *number*, in some degree, compensates for *magnitude*. Thus, instead of the single stupendous tuft of the palm, we have the numerous congregated buds of our deciduous trees ; instead of the gigantic and solitary grasses of the torrid zone, we have the smaller and gregarious varieties. Some of these varieties, as the *Cerealia*, or Corn tribe, with their myriads of seeds, give us an inexhaustible supply of farinaceous aliment ; while others, as the Grasses properly so called, clothe our meadows with verdure, even to extreme latitudes ; and are equally productive of matter purely herbaceous. In the warmer parts of the temperate zone, the Olive and the Vine afford the oleaginous and the saccharine principles, under a form, different, but not less useful than the oil and the sugar of the tropics ; while in the colder parts, various seeds, and hardy fruits, produce an ample store of the same valuable articles, though in a condition still further modified.

In the preceding sketch we have intentionally kept out of view the existence of animals, that we might here ask the question,—Of what use is all this amazing exuberance of superfluous matter throughout the globe? The adaptation of plants to the climates in which they flourish, is evidently the work of an intelligent Creator. But can this apparent waste of materials, and of labour, be reconciled with the same wise

agency? Surely, the mere existence of vegetation did not require such prodigality. Seeds,
for instance, need not have been enveloped in
bulky fruits; nor need they have been produced
by myriads: and all that foliage, all those flowers,
and roots, in such amazing profusion, of what
use are they; why were they so created? If we
regard vegetation as a thing simply adapted to
climate, and existing for its own sake alone, the
question scarcely admits of a rational answer.
But if we consider at the same time, the existence
of animals, and view these superfluities as the
means by which animal existence is principally
upheld; every difficulty vanishes, and the splendid design of the whole wonderful scheme becomes at once apparent. We are thus brought
to the consideration of animal existence.

SECTION II.

Of the Distribution of Animals over the Earth.

ANIMALS have been so constituted, that food is
to them indispensable : they can, therefore, exist
only where their food has been supplied by nature. On land, at least in the warm and temperate climates, by far the greater proportion of
animals derive their subsistence, either directly
or indirectly, from the vegetable kingdom. For

those animals which are themselves carnivorous, prey on vegetable feeders much oftener than otherwise ; and are thus remotely dependent on vegetables. Of the habits of animals living in the sea, and thus concealed from our view, we know still less : but in general, they appear to prey on each other ; the stronger, as is usual, devouring the weaker.

We have seen the wonderful diversities prevailing among vegetables, in different situations and climates ; and it may be truly said, that the diversities among animals are not less numerous, and are even more extraordinary. Thus, in every climate and soil, almost every herb has its appropriate inhabitant ; some little being, that comes into existence, passes its ephemeral life, and dies on the same plant ; to which, therefore, that plant constitutes the world. Nay, in general, even different parts of the same plant have each its separate occupants, one feeding on its fruit, another on its flowers, a third on its leaves, perhaps a fourth on its very woody core. This almost infinite diversity, and infinity of number, are principally confined to the smaller animals, or insects. As animals increase in size, the number of species as well as of individuals constantly diminishes. Thus, while there are hundreds of species of the Beetle tribe, and the individuals are countless, there may be considered to be only one Elephant ; and while

Shrimps are in numbers like the sand on the sea-shore, the Whale is as much a solitary species. This striking difference with regard to numbers, has been considered to arise necessarily from a law of nature; and in one respect such an explanation is very obvious : but, in another point of view, we may contemplate an admirable evidence of design. It is clear that millions of elephants could not exist; if for no other reason, from want of food: but why should millions of beetles exist? why should these little creatures,—whose life is so transitory, that it consists of little more than of being born, and of dying ; whose structure is so frail as to be liable to be annihilated by the slightest accident ; who are everywhere surrounded by all sorts of enemies, to many of which they constitute a natural prey ;—why, we ask, in spite of all these obstacles and dangers, should these insignificant animals contrive to exist in the numbers we see? No natural law, surely, will explain the appearance of such multitudes. The difficulty requires another solution ; and the only solution which can be offered is *design*—that it was so designed by the Great Author of nature. And how has He effected His purpose of multiplying to such an extent these little animals? The answer is, simply, by increasing their fecundity. Had beetles, like elephants, brought forth only one young at a time ; long ere now, their race would have been

P.   C C

exterminated: but, being produced by thousands; some of the numerous offspring chance to escape, and thus the race is perpetuated.

We shall not dilate further on the arrangements which have been made for the existence and preservation of animals, but shall proceed to consider their distribution.

The distribution of animals over the earth may be conveniently treated of under the same heads as the distribution of vegetables; and, first:—

1. *Of the Differences existing among Animals that live in similar Situations in different Parts of the World.*—The dwelling of animals in the waters is, perhaps, the most remarkable as regards their localities. Now, since, from circumstances formerly stated, the distribution of temperature is very different in the sea from what it is on land; and since most aquatic animals prey on each other, and consequently in some degree are independent of climate; the distribution of such animals over the globe, follows laws materially different from those which regulate the distribution of land animals. This distribution of temperature, more especially affects the distribution of animals in high latitudes; and must be taken into account at the very outset of this part of our enquiry. We shall, therefore, state concisely the distribution

of land animals, and of sea animals, apart from each other.

The distribution of land animals, resembles to a certain extent, the distribution of vegetables: for though animals differ from plants, in being endowed with the power of locomotion; yet, as the larger number of animals are dependent on vegetables for their subsistence, they are necessarily confined to those places where their peculiar food may be obtained. This limitation of range, is most observable in the case of the smaller animals. The existence of many kinds of insects, especially, is intimately connected with the existence of certain plants. In every tribe of animals, however, there are species that occupy very different localities. Thus, in the same tribe, some species dwell on the mountains, others on the plains; some species are most numerous on the sea-coast, others live on trees, while others of the same tribe burrow in the ground. All these diversities, in regard to residence, are probably influenced, like many others, by the greater or less degree in which the locality affords the means of obtaining subsistence. But, in many animals, there is also a wonderful adaptation of structure to the place they inhabit; proving, beyond a doubt, that the distribution of animals has been arranged by design; and that they form but a part of the great symmetrical whole of creation.

In animals that dwell in the water, the same peculiarities of habitude are observable, as in those that dwell on the land. Thus, it is perfectly known that many animals can live only in salt water; others only in fresh. Some prefer the deep and open sea, others are met with only in shallow water, or at the mouths of rivers. Of those that flock to the coast, some shun turbid water, others burrow in the mud. In short, though the habits and adaptations of aquatic animals can be less satisfactorily ascertained; there is every reason to believe, that they are at least as wonderful, as the habits and adaptations of the occupants of the land.

There is an equally striking diversity in the animals, as in the plants, of similar localities and climates in different parts of the world. Thus, in the old world, though many genera exist, common to the analogous climates, on the north, and on the south, of the equator; yet the species are totally different. For instance, the northern hemisphere possesses the *Horse*, and the *Ass;* while, in the south, these species are represented by the *Zebra* and the *Quagga*. In the southern hemisphere, there also exist many species which are quite peculiar; as the *Giraffe*, the *Cape Buffalo*, and a variety of animals having the *Antelope* form. So, likewise, the animals of the old and those of the new world are, in general, quite distinct;

unless, perhaps, towards the north, where the
two continents approximate; and where, in consequence, there are some species common to
both. Thus the *Elephant*, the *Rhinoceros*, the
Hippopotamus, the *Giraffe*, the *Camel*, the *Dro-
medary*, the *Horse* and the *Ass*; also the *Lion*
and the *Tiger*, and various species of *Apes*,
Baboons, and other animals, with which we are
familiar in the old world, were not found in America. On the other hand, the American species,
the *Lama*, and the *Peccari*; and among carnivorous animals, the *Jaguar*, or American tiger;
also the *Agouti*, the *Paca*, the *Coati*, the *Sloth*,·
and others, were equally unknown in the old
world. Again, the animals of New Holland differ, like its vegetation, not only from all those
of our continent, but from those of all the world
besides. In New Holland, there are more than
forty species of marsupial or pouched animals,
of which the *Kangaroo* is that with which we
have become best acquainted; while every where
else, there is hardly a known instance of a
pouched animal. Nor are these differences confined to the more perfect animals. They are
even more striking as we descend in the zoölo-
gical scale. Thus among birds; the individual
species of the *Parrot* tribe, that are found in
America, differ altogether from those of Africa;
and those of Asia differ from both. The minute
and beautiful family of *Humming birds* is pecu-

liar to America. While the species of the common *Grouse* of this country is met with in no other part of the known world.

From the class of reptiles, may be mentioned the Great Saurian, or Lizard tribe. Thus the *Crocodile* of the Nile, is entirely different from the *Cayman* of America; and even from the *Gavial* of the Ganges. In the division of snakes, too, the *Boa* of India differs from the nearly allied *Python* of America; and of the poisonous varieties, the *Rattlesnake* is peculiar to America, the *Cerastes* to Africa, the *Hooded snake* to Asia. As we have already stated; the diversities among insects are still more numerous and remarkable than among the larger animals. To enter into details would be endless; we shall therefore mention only one of the best known and widest extended of all the insect tribe, viz. our *common Bee*. This insect did not exist in America, or in New Holland; though it is found in all parts of the old world; the wax and honey of Europe, Asia, and Africa being obtained from species having a close resemblance to each other.

Nor are these differences confined to land animals; the inhabitants of the waters are equally diversified. Thus the *Whales* of the northern ocean are quite unlike those of the south; as are the *Seals*, and other analogous animals in the polar regions. Different seas, also, not only when far apart; but even some which freely communicate, are often exceedingly dissimilar. Thus, the fishes of the Arabian Gulf are said to differ entirely from the fishes of the Mediterranean; notwithstanding the proximity of these seas. Nor does the remark apply only to the larger fish in these seas, but holds equally with respect to their testaceous and molluscous species.

Such are a few of the more striking facts with regard to the distribution of animals in similar climates and localities throughout the world. We now shall briefly speak,

2. *Of the Effects of Diversity of Climate on the Distribution of Animals.*—In tropical climates, the qualities of animals, as well as those of vegetables, are developed to the utmost; whence arises that harmonious adaptation of all the works of nature, conspicuous, indeed, in all climates, but in Tropical climates more especially. For, where else than amidst the profuse exuberance of the vegetation within the tropics, could the *Elephant*, the *Rhinoceros*, the *Giraffe*, and other large quadrupeds find subsistence? Where else could we expect to see such birds as the *Ostrich* and the *Cassowary?* such reptiles as the *Crocodile?* such serpents as the *Boa?* To what other part of the world could the magnificent butterflies, the enormous beetles and spiders, be so appropriate. Even

among the marine animals of Tropical climates, there is the same wonderful enlargement of size. Thus certain species of the *Crab* and *Lobster*, and various shell fish, often attain an enormous magnitude. Nor is there a developement of size only, but of every other property in an equal degree. Countries within the tropics exhibit the most beautiful forms—the most splendid colours in nature. There, in short, is the most astonishing display of all those things which seem to be entirely ornamental: as, for example, the singular plumage of the *Birds of Paradise*—the gaudy liveries of many of the *Parrot* tribe—the extraordinary and diversified forms and colours of many insects and shells.

Not only is there in Tropical climates an assemblage of all the concomitants of productiveness, and utility, and embellishment of every kind; in these climates, there is another, and not less marked demonstration, of the power and the wisdom of the great Creator. Within the Tropics *death*, the last, the inevitable scene, assumes a character as new and diversified as that of the life it terminates. There, rages the implacable ferocity of the Tiger, and of the larger beasts of prey; there, the fangs of the serpent are charged with the most malignant venom; there, even the insects are as formidable as they are numerous. Nor is this intensity of the destroying power, incongruous or without an object; but evidently is in perfect harmony with the rest of creation, and with the design of the Creator. The wonderful productiveness of animals in Tropical countries, renders unavoidable, some checks against their excessive increase: and in devising these, the great Author of Nature has displayed the same attributes which are manifest in all his other works. No one who seriously reflects, can doubt either the wisdom or the benevolence of the provision. For why are Tigers and Serpents confined to those parts of the world, where their existence is not only accordant, but where, for one great purpose at least, they are even necessary. Surely such limitation could have happened only from design; and the argument is strengthened a hundred fold, when we contemplate the striking display of wisdom and of power, exemplified in the singular adaptation of the structure of these animals, to their peculiar habits. Thus how wonderfully is the Tiger formed: how extraordinary as well as wonderful, must be the organization of Serpents! Who (unless he had witnessed the fact) could have believed, that the animal frame was capable of separating from its juices, and of retaining with impunity, a poison instantly fatal, not only to other animals, but to the animal itself; if again mingled with the juices from which it had been separated!

Nor in all these things is the benevolence of the Deity less conspicuous, than his wisdom. *All must die;* and death from rapacious or venemous animals, is probably not in any degree more painful, than many other modes of death which we constantly witness. There is, in truth, to our own narrow and selfish feelings, something exceedingly painful in the idea of being torn to pieces by a Tiger, or stung to death by a Rattlesnake; but how many thousands of little mice are destroyed by cats? and how many myriads of unfortunate flies are poisoned by spiders, every day we live? and yet we hardly commiserate them. The question, therefore, is simply a question of degree: and viewing the existence and the destruction of animals, as they ought to be viewed, on the great scale, we find that the whole is perfectly in unison. While in temperate climates we have cats and spiders, designed as checks on over productiveness; amidst the grandeur and the luxurious developement of the Tropics, the same wise purpose is executed by the Tiger and by the Rattlesnake.

As we advance from the Equator into the temperate climates; the size of animals in general, like the size of vegetables, becomes gradually smaller. Like the vegetables, too, the animals of temperate climates are more gregarious than within the Tropics. Hence *number*, as among vegetables, compensates in some degree for

diminished *magnitude*. The two kingdoms of
nature therefore are beautifully analogous ; for
the gregarious grasses, which, as we before observed, form so marked a feature in the vegetation of temperate climates, constitute in one
shape or other the principal food of the gregarious tribes of animals. Thus the whole cattle
tribe—The *Ox*, the *Sheep*, the *Goat;* the different varieties of *Deer;* the *Rabbit* and *Hare;*
also the *Horse* and the *Ass;* with a multitude of
other well-known animals, of a similar character,
are natives chiefly of temperate climates ; and
obtain their nourishment almost entirely from
the grasses. Among birds, the numerous species of the Gallinaceous, or Fowl tribe, may be
said to derive their food from the same source.
As regards the existence of animals, therefore,
the gramineous tribe of plants is more important
than perhaps any other ; and could not be annihilated, without the destruction of the present
order of living beings.

As further examples of animal species indigenous to temperate climates, may be mentioned
the *Canine* species and those allied to it, most
of which are more or less carnivorous ; also the
Hog; and a variety of other animals that need
not be here enumerated. The Hog tribe, as is
well known, are omnivorous ; but in their natural
state, they feed principally on the seeds and
roots of plants. Among birds peculiar to tem-

perate climates, are various tribes of *Water-fowl* that subsist on fish and on insects. Of the smaller land birds, the various *Songsters* offer a remarkable contrast to the birds of similar form within the Tropics; not only from their more melodious notes, but from the simple colouring of their feathers. In temperate countries the *Insects* are still exceedingly multiplied; though, in general, like the other animals, they are much smaller in size than the Tropical insects; their forms, their colours, and other peculiarities, are, also, much less remarkable.

As we advance toward the Poles; the animals of temperate climates are observed gradually to decline in number. The vegetable feeders become reduced to a few hardy species; and at length in the remote north and south scarcely any vegetable feeders remain. So far as shrubby plants continue to grow in these inhospitable regions, individuals of the *Squirrel* tribe find subsistence on their seeds and roots. But the most remarkable herbivorous animal is the *Reindeer;* whose principal food is afforded by nature, in a species of moss peculiar to very cold climates. Those animals which exist beyond, are either carnivorous or piscivorous. The *Arctic Fox* and the *Bear* are familiar instances, as terminating the Zoological series, viewed in connection with the influence of climate.

We have, in the last place, to notice what is

most remarkable in the distribution of *Marine* animals.

For the reasons before stated, the general temperature of the ocean, differs considerably from that of the land. Owing to this difference of temperature, and to the peculiar mode of subsistence of marine animals, which is obtained chiefly from the waters they inhabit ; the distribution of these animals varies much, as compared with the distribution of animals that are entirely terrestrial ; particularly within the frigid zone. It is true, indeed, that in all climates, the denizens of peculiar localities, as fresh water species, and species which resort to the shallows on the coast, are influenced by the climate nearly as much as land animals : and within the Tropics, this influence extends in some degree even to the species that dwell on the wide ocean. But far to the north, and to the south, such species are influenced in a manner altogether different. Thus the largest of known animals, the *Whale*, and of course those other animals which become its prey, roam through the utmost-Polar seas ; where on land the intensity of the cold would prevent the existence of any animal whatever. The whale is enabled to live in so rigorous a climate, solely in consequence of the greater warmth of the Polar ocean, formerly explained. Among the larger inhabitants of the ocean in Tropical climates, may be mentioned the *Shark* tribe; which in respect of ferocity and voraciousness, may be classed with the tiger, or any kindred species on land. The influence of climate on marine animals is further shown, as we have said, by the enormous size of many of the Tropical shell-fish and mollusca. The colouring of these and also of other productions of the Equatorial seas, often exhibits so much lustre and beauty, as to rival the most splendid of the feathered race. In temperate climates, and from the equal temperature of the sea, even within the frigid zone, it is remarkable that fish, like terrestrial animals, are much disposed to be gregarious. The shoals of *Herring*, *Mackerel*, and other well known visitants of our coasts, are familiar examples of the gregarious tendency. The *Salmon* and the *Sturgeon* may be adduced as instances of fish inhabiting chiefly the rivers of the temperate and colder countries. While in the same climates, instead of the magnificent *Pearl oyster* of the Tropics, there appears our common *Oyster*, so diminutive and unsightly, yet so profitable to man.

We have thus seen that animals, like plants, have in general been adapted to particular climates. The numerous cold-blooded animals of the Tropics—even the warm-blooded Tiger itself, amid the Polar snows, would instantly perish. The Arctic bear would be not less unable to live, under the scorching rays of a

vertical sun. Yet though adaptation to one climate be the general law regarding animals as
well as plants; some species of animals have,
as remarkably as some species of plants, the
faculty of accommodating themselves to all climates. These species, like the plants similarly
endowed, are for the most part natives of temperate climates; the transition from such climates to either extreme, being much less violent
than from one extreme to the other. Thus our
domestic animals have been successively introduced into the New World, at various periods
since its discovery; and are now, in incredible
numbers, spread over the whole of that vast continent, from Canada to Paraguay. The greatest
increase has been of the Horse, the Ox, the
Sheep, the Goat, the Dog, the Cat, and the
Hog. The Rat, too, though an unwelcome in-
truder, has been not the least prolific. The different varieties of domestic Poultry have multiplied to an equal extent. Even Insects have
been introduced, and widely spread, as is well
known to horticulturists.

Like plants, most animals also are readily
domesticated, and thrive in climates similar to
those of which they are natives. The most
striking instance is the Rein-deer; so lately as
in the year 1773 introduced into Iceland, and
now exceedingly numerous in the interior of
that country. From these powers of accommo-

dation to climate, from the agency of man, and from accidental causes; the distribution of the larger animals over the globe has, in comparatively recent times, been very much modified. Nor is there any reason to believe, that the distribution of these animals is yet stationary; but, on the contrary, that their distribution will undergo still further changes.

Among the more remarkable habits of animals, may be noticed the *migratory* propensities of certain species. The migration of land animals, is always much limited, and may be entirely prevented by natural obstacles; such as the asperities of the earth's surface; sands; deep rivers, or other large accumulations of water. But many birds and even insects, possessing powerful locomotion, and whose course is through the air, may literally be said to follow the sun in their migratory progress. It is hardly necessary to state, as examples, the birds of passage, so well known as the Swallow and the Cuckoo. These birds during the summer months visit our northern climate, and feed on insects, whose multiplication would otherwise be boundless. Having fulfilled their office here; on the declination of the sun, they again retire to the south; and are succeeded by different birds from countries still further north. Such are the Woodcock and others, which escape to our shores from the rigorous cold of a Polar winter. Nor is migration confined to the higher classes of animals. The wonderful powers of flight possessed by many insects, enable them to travel over an immense extent of country. The Locust and the Ant tribe are familiar examples. These insects occasionally migrate in countless swarms from the lands to which they are indigenous, and lay waste others far remote.

Equally remarkable is that habit of animals termed *Hybernation*. Like the plants of temperate climates, some animals have the faculty of passing the colder season of the year in a state of sleep. The Hedgehog and the Dormouse may be mentioned as examples of quadrupeds possessing this faculty. Additional instances might be given in all the classes of animals. Nearly allied to Hybernation, is that remarkable instinct which guides many of the inferior animals to deposit their eggs in the earth, or in some other place of safety ; that they may be preserved during the season of diminished temperature. This instinct is particularly observable in insects whose lives are ephemeral, or are, at the utmost, prolonged for a summer.

There is yet another circumstance that remains to be noticed, as being connected with the adaptation of animals to the climates in which they live ; namely, the *Clothing* or covering with which animals have been supplied by

P. D D

nature. Every one is acquainted with the general fact, that wool, fur, eider-down, and similar articles, are obtained for the most part, not from the copious source of every superfluous production, the countries within the tropics; but from the cold, and comparatively unprolific regions of the temperate and of the frozen zones; where they have constituted the appropriate vesture of different animals. Perhaps, in the whole range of creation, there is not any thing more calculated to excite our admiration. However we may view these means of guarding animals from being injured by the cold; whether as a part of that conservative faculty with which animals have been endowed, and by which their existence is maintained; or as an immediate act of Providence; still the adaptations are so striking and obvious, as to render it impossible to doubt for a moment, that they have all been contrived for the purpose which is accomplished; and that they are the results of fore-knowledge and of design.

We have thus given a rapid sketch of the distribution of animals over the globe. In this sketch, we have endeavoured to point out the wonderful adaptations of the several classes of animals to the circumstances in which they are placed; together with the beautiful symmetry and equilibrium, exhibited in zoology, not less than in the arrangements of inanimate matter.

Throughout, we have intentionally, and as far as was possible, avoided those details, the consideration of which belongs to other departments. But it has been our aim to state such prominent facts, as appeared best calculated for the elucidation of our argument. In particular, it has been our desire to show—how number among the weak is made to compensate for magnitude among the strong; how exuberance in one species is made to contribute to the existence of another; how ornament and boundless profusion characterize the countries within the tropics, while the temperate climates are not less distinguished by utility and capacity for change; how, even in the rigorous and barren neighbourhood of the Poles, where life becomes a struggle for existence, animals have been expressly furnished with clothing appropriate to these regions;— in short, we have endeavoured to explain, how every animal, in every climate, has its day; and by some peculiar contrivance, has been enabled to maintain its rank in creation, and to assist in preserving the general equilibrium.

Hitherto we have considered the works of nature without reference to *Man*. For aught we can see to the contrary, they might all have existed, and every arrangement and operation might have been very nearly, if not exactly, the same as at present; though man had never been called into being. But still, for a moment

longer, keeping man's existence out of view;
let us, as under a former division of this Treatise, enquire, what would have been the *use* of all
this elaborate design, without an ulterior object.
Would an intelligent Creator have made such
a world, and have left it thus incomplete? It
is evident that the other beings inhabiting this
earth, live and die, without in the slightest degree comprehending the vast system of which
they constitute a part. Hence they are merely
unconscious agents, from which their Maker,
while he has furnished them with the instincts
necessary to their existence, and has awarded
equal justice to all, has yet chosen to withhold
the privilege of reason. That a Creator, evidently as benevolent as he is wise, might, for
his own gratification, have made such a world,
and without any other inhabitants, is indeed
possible. But, even admitting that possibility,
the probability surely is, that he would not
there have finally "rested from his labour."
His benevolence would have prompted him to
communicate to other beings, a portion of the
gratification, which he himself is supposed to
derive from the contemplation of his works.
In the beautiful world which he had created,
He would have wished to see *one* being at least,
capable of appreciating to a certain extent his
design and his objects. Such is a plain inference deducible from the manifest attributes of the

Creator; and what is the fact? Is not man such a being as we have supposed? Throughout the world, though perfectly independent of him, is there not a clear foretoken of his existence? Has he not been placed at the head of that world, so obviously prepared for him; and thus constituted " the Minister and Interpreter of nature?" Surely no one will be inclined to doubt that such is the position of man with reference to other animals. Equally undeniable, is the striking accordance of these deductions from the view of external objects, with what is written of the origin of man by the sacred historian: " and God said, that it (the world which he had prepared) was good. And God said, let us make man in our own image, after our own likeness, (that is to say, endowed with reason and with the power of reflection). And let him have dominion over the fish of the sea, and over the fowl of the air, and over the cattle, and over every creeping thing, that creepeth on the earth."

We thus arrive at another, and to us the final step in the great design of the Omnipotent: the creation and the faculties of Man.

Section III.

Of the present Position and future Prospects of Man.

The consideration of the faculties of man, and of his position in the world he inhabits, belongs, in all its details, to another department. We advert to these subjects here, with the view only of completing our sketch of the physical relations of animated beings. The observations we have to offer will be comprised under two heads:—as to the means, by which man has acquired and maintains the *ascendency* he enjoys:—as to the conclusions to be drawn, from man's *elevated position*, and from his superior *intellectual character*.

With regard to the means by which man has acquired and maintains his *ascendency*, it may be observed, that these means are quite peculiar; and far from being such, as at first, perhaps, we might deem conducive to such an object: though when once known and understood, the beautiful design and harmony they evince, immediately become apparent.

The supremacy of man has *not* been the result of his own personal strength; nor is it so upheld. On the contrary, many animals are

larger and more powerful than he is ; while few
of his size, are naturally so incapable of self-
defence ; or during so long a period suffer from
the dependent helplessness of infancy, and of
old age. Neither is his frame superior in external adaptation to climate : for while nature
has furnished other animals with clothing appropriate to the temperature in which they live,
man has been brought into being absolutely
naked ; and moreover remains so, in every
climate he inhabits, from the Equator to the
Poles. Lastly, the pre-eminence of man has not
been owing to his more extensive range of diet ;
or to his greater ability for assimilation : for
though man be omnivorous in one sense of the
term, he is not omnivorous according to the
application of the term to other animals ; that is
to say, man does not eat indiscriminately of
every kind of aliment, in the state in which it is
afforded by nature ; for even in his rudest con-
tion, he adopts some process of cookery. How
then has man gained the high station which he
occupies? The answer is simply — by his
Reason. Man has been created a reasonable
being ; and this endowment amply compensates
to him for the want of the animal requisites of
strength—for deficiency of natural covering—
and for his restricted ability in assimilating his
food. By his reason, he is enabled to command
the strength of the elephant; to choose from

every production of nature whatever is adapted for his clothing, and thus to array himself according to his pleasure, or the exigencies of the climate in which he resides; to extract wholesome nourishment from the most unpromising, even from the most deleterious articles. There was no necessity, therefore, why man should himself be as unwieldy as an elephant; or be encumbered with any vesture that in some situations might be oppressive; or be able to digest, without culinary preparation, any coarse and intractable substances. Thus, mere animal endowments not being requisite; the Creator's wisdom has been displayed in another manner, and with a wider scope. In furtherance of his design, He has limited the bulk of the human species to that happy medium, combining strength with convenience; and to an organization delicate and sensitive in the highest degree, but nevertheless accommodating, He has superadded a form at once peculiar, appropriate, and beautiful!

When speaking of temperate climates, we remarked, that they seemed to be characterized by the utility of their productions; and that the plants and animals of these climates, generally possessed greater powers of accommodation than those of either of the extreme climates. Now Man, by an express arrangement of his Maker, has apparently been constituted a native of temperate climates; and only in these climates can his powers be said to be completely developed. Within the tropics, indeed, human existence is flourishing; for there the immediate bounty of Providence affords to man a copious and admirably adapted nutriment. Yet in the midst of that profusion, and without any adequate motive to call forth exertion, his reason too often languishes; while his animal tendencies predominate; and his life is spent in apathy and in sensual gratifications. On the other hand, under the cheerless sky of the frigid zone, imperfectly nourished by scanty and unsuitable food, the powers of his mind, like those of his body, are stunted; or are engaged solely in combatting the rigours of his situation. But in the temperate climates, the evil consequences of both these extremes are avoided, while the beneficial influences of climate remain. Urged by the stimulus of necessity, and at the same time having at his command the astonishing capability of nature, man is, in temperate climates, surrounded by motives of every kind; and his faculties thus attain their utmost developement. As familiar examples of the effect of this expansion of the human reason, let us view man under the three aspects to which we have before alluded; namely, with reference to his *strength;* his *food;* and his *clothing*, inclusive of his *habitation.*

In the first place, with regard to his strength. The strength of man is not only that which is his own, almost infinitely magnified by ingenious mechanical devices of every kind, and of every degree, up to the stupendous agency of steam; man has, moreover, subdued to his service many of the larger animals, while those he cannot so appropriate, he destroys. As weapons, he wields every instrument offensive and defensive, from the rude but effective club or arrow, to the warlike engines to which he has applied the discovery of gunpowder. Whatever his wants require, he obtains by tools; from the humble spade, to that perfection of machinery, which almost rivals the operations of intelligence itself. In the next place, view man with reference to his food: what wonders has not his reason enabled him to achieve among the fellow inhabitants of his own temperate climate. In the vegetable kingdom, let us consider the astonishing mutations and increase of the *cerealia*, or corn tribes; the transformation of the sour and forbidding Crab into the rich and fragrant Apple; of the harsh and astringent Sloe into the delicious Plum; of the coarse and bitter sea side *Brassica* into the nutritious and grateful Cauliflower: all which changes, and numerous others of a like kind, have been effected by man. Nor have the transformations which he has produced among animals been less wonderful than those among vegetables. All the numerous varieties of cattle, of sheep, of horses, of dogs, of poultry, and of all the other animals reared as food, or for any purpose domesticated, have sprung from a few wild and unattractive species; and have been made what they are, in a great degree, by his intervention. Moreover, the most useful of these varieties of animals have been transported by man into every region of the globe, to which he has himself been able to penetrate. Lastly, in the clothing and habitations of man, the surpassing influence of his reason is equally conspicuous. For covering his naked body, a surface of considerable extent is necessary; larger, indeed, than is presented by any natural texture, unless, perhaps, by the skins of other animals, or by the leaves of some plants; which therefore, in the rudest states of society, usually constitute his only dress. But by the art of *weaving*, he has been enabled to produce garments of any size, and from materials which would seem the least fitted for such conversion. Thus man can not only clothe himself in any manner, and according to the temperature of the climate in which he lives; but he can associate with the articles of his dress every species of ornament his fancy may dictate. His choice of materials for the construction of dwellings is not less extensive than that of his clothing. As climate and other circumstances may require, he abides in the humble cabin; or in the splendid palace: in the temporary hut; or in the enduring castle, formed to withstand alike the tempest of war, and of the elements.

Such is man, and such are a few of those great changes in this world, which, under the guidance of his reason, he has had the power to accomplish. And what a splendid evidence of design, and of preconcerted arrangement on the part of the great Creator, is thus exhibited, by viewing the inherent properties of matter, and its various conditions, with reference to the works of man. Had water, for instance, not been constituted as it is; man could never have formed the steam engine. Had not the productions of the temperate climates been formed with that capability for change, by which they are so much distinguished; man could never have so moulded them to his uses, by altering their character. There was no reason why such properties should have been communicated; there was even no reason why the objects in which these properties exist, should have been created. But they have been so created; and what are we to infer? No one surely will now maintain, that the objects of nature possessing these properties, have been the result of chance, or have been created without an end. They must therefore have been created with design; and if with

design—most obviously with design having reference to the being man, *not yet in existence*.

Thus far we have considered the state at which
the earth has arrived, and man, an animal endowed with reason, placed as its chief inhabitant. But we may yet extend our view to the
prospects in *futurity*.

We have seen that this earth has not suddenly
emerged from chaos to its present condition ;
but that by a succession of violent and disruptive changes, it has been progressively brought
into different conditions, and progressively te-
nanted by higher orders of beings. We, the
last of the series, in our own creation and in the
faculties with which we have been endowed,
behold the most striking exemplification of the
wisdom, and of the power of the Deity. But
does the great design abruptly terminate here?
Has this earth arrived at the ultimate stage of its
existence? Have its inhabitants attained the
utmost perfection of which they are capable?
Are there not further convulsions, and still higher
orders of beings in contemplation? The answers
to these questions are known only to the great
Author of the universe, and concern us not.
There is one question, however, connected with
this subject, in which we are deeply and per-
sonally interested—*What is to become of man?*
Is the being who, surveying nature, recognises

to a certain extent, the great scheme of the universe; but who sees infinitely more which he does *not* comprehend, and which he ardently desires to know;—is he to perish like a mere brute—all his knowledge useless; all his most earnest wishes ungratified? How are we to reconcile such a fate with the wisdom—the goodness,—the impartial justice—so strikingly displayed throughout the world by its Creator? Is it consistent with any one of these attributes, thus to raise hopes in a dependent being, which are never to be realized? thus to lift, as it were, a corner of the veil—to show this being a glimpse of the splendour beyond—and after all, to annihilate him? With the character and attributes of the benevolent Author of the universe, as deduced from His works, such conceptions are absolutely incompatible. The question then recurs—What is to become of man? That he is mortal, like other animals, sad experience teaches him; but does he, like them, die *entirely?* Is there no part of him, that, surviving the general wreck, is reserved for a higher destiny? Can that, within man, which reasons like his immortal Creator—which sees and acknowledges His wisdom, and approves of His designs, be mortal like the rest? Is it probable, nay, is it possible, that what can thus comprehend the operations of an immortal Agent, *is not itself immortal?*

Thus has reasoned man in all ages; and his desires and his feelings, his hopes and his fears, have all conspired with his reason, to strengthen the conviction, that there is something within him which *cannot die.* That he is destined, in short, for a future state of existence, where his nature will be exalted, and his knowledge perfected; and where the GREAT DESIGN of his Creator, commenced and left imperfect here below, WILL BE COMPLETED.

BOOK III.

OF THE CHEMISTRY OF ORGANIZATION:

COMPREHENDING A SKETCH OF THE CHEMICAL PROCESS OF DIGESTION; AND OF THE SUBSEQUENT PROCESSES BY WHICH VARIOUS ALIMENTARY SUBSTANCES ARE ASSIMILATED TO, AND BECOME COMPONENT PARTS OF, A LIVING BODY.

HAVING in the foregoing pages, given a summary view of the Chemical properties of bodies not organized, and of the laws of their union; having also considered the general relations of inanimate matter and of organized beings, on the great scale in which they are offered to us by nature, together with the present position and future prospects of man; we now proceed, in the last place, to enquire more particularly into the means by which organization is accomplished; or, in other words, to give a summary view of those chemical properties, and laws of union, by which organized beings are distinguished from inorganic matters.

CHAPTER I.

OF THE NATURE AND COMPOSITION OF ORGANIZED
BODIES IN GENERAL, AS COMPARED WITH INOR-
GANIC MATTERS.

" A LIVING being considered as an object of chemical research, is a laboratory, within which a
number of chemical operations are conducted;
of these operations, one chief object is to produce
all those phenomena, which taken collectively
are denominated *Life;* while another chief object is to develope gradually the corporeal machine or Laboratory itself, from its existence in
the condition of an atom, as it were, to its utmost
state of perfection. From this point of utmost
perfection, the whole begins to decline as gradually as it had been developed; the operations
are performed in a manner less and less perfect,
till at length the being ceases to live; and the
elements of which it is composed, again set free,
obey the general laws of inorganic nature."*
Such is the history of organic existence; nor,
though the periods of developement and of decay

* Berzelius Traité de Chimie, tom. v. p. 1.

P.   E E

be infinitely varied in different species, does a single individual remain for a moment stationary ; but all, sooner or later, transcend their prime, and finally share the common lot of dissolution.

That peculiar principle or principles, which under some condition or other, exists in all organized beings, and by which they are distinguished from inanimate matter, has received various appellations. In the present enquiry these principles may be viewed as agents ; and to discriminate them from Heat, Electricity, and other agents operating on inorganic matters, they may be denominated *organic agents*. In conducting our investigations into the nature of these principles or agents, our difficulty will be much lessened, by endeavouring previously to have a clear understanding of what these agents actually do. We shall, therefore, in the first place, give a short sketch,

1. *Of Organic Bodies considered as Chemical Compounds.*—In their well-marked forms, no two things perhaps can be conceived to offer a stronger contrast, than the two great divisions of organic bodies—vegetables and animals. Yet these two kinds of bodies so gradually approximate, and seem even to coalesce, that it is not possible to say where the one ends and the other begins. The same remark applies to the chemical composition of vegetables and animals.

Vegetable substances, in general, contain essentially no more than three elements, *Hydrogen*, *Carbon*, and *Oxygen*; while animal substances usually involve a fourth, *Azote*. Yet there are many vegetable substances, of whose composition, azote forms a considerable part; while certain animal substances are entirely wanting in that principle. It is obvious, therefore, that the mere chemical composition of a substance, at least its essentially consisting of three or of four of these elements, will not enable us to determine whether it be vegetable or animal; and that, in many instances, when this point happens to be doubtful or unknown, we must have other data before we can form a conclusion. Besides the four constituent elements, of which all organic substances are essentially compounds; other principles generally enter into their composition. These other principles are in very minute quantity, and are not so essential to the existence of organic substances, as the four constituent elements above named; yet, however minute the quantity, the influence of these other principles seems to be most important; they are, *Sulfur*, *Phosphorus*, *Chlorine*, *Fluorine*, *Iron*, *Potassium*, *Sodium*, *Calcium*, *Magnesium*, and probably more besides. These principles have, by most chemists, been deemed extraneous, or foreign to organized bodies; but we shall presently show, that there is good reason to believe, that the office of such additional prin-

ciples, though different from that of the four constituent elements, is nevertheless most remarkable.
These four elements, along with the additional
principles, are, in the present state of our knowledge, alike denominated, *The Ultimate Elements*
of organized bodies; but hydrogen, carbon,
oxygen, and azote, may be termed, for sake of
distinction, the *essential* elements; and sulfur,
phosphorus, &c. the *incidental* elements of such
bodies.   The combinations of these ultimate
elements with one another, according to certain
laws, produce what are denominated the *Immediate*, or *Proximate Elements* of organized bodies.
Of such proximate elements, *Sugar*, *Oil*, *Albumen*, &c. are familiar examples.

Perhaps it may be stated as a general law,
that no substance, entering into the composition
of a *living* plant or animal, is so pure as to be
capable of assuming a regularly crystallized
form.   Instead, therefore, of being defined by
straight lines and angles, almost all solid organized substances are more or less rounded, and
their intimate structure is any thing but crystallized.   The composition of organized fluids is
equally heterogeneous; and though the basis of
nearly every one of such fluids be water, many
of them contain a variety of other matters.

Organized bodies may be ranged under two
general classes; those which though they do not
crystallize, while in the living plant or animal,
can yet, by various processes, be so far separated

from extraneous matters, as to be obtained in a
state of purity, and thus be made to assume the
crystallized form ; and those which cannot under
any circumstances be made to crystallize.   The
first substance of the crystallizable class which
we shall notice, is *Sugar*.

Sugar has been ascertained, and is now generally admitted, to consist of three essential elementary principles—hydrogen, oxygen, and carbon ; it is besides remarkable, that the hydrogen
and the oxygen in sugar, are exactly in the proportion to each other, in which they form water.
It has been, therefore,. with great probability
inferred, that these two elements are really so
associated in sugar ; consequently, that sugar
is a compound of water and carbon ; or, in the
language of Chemists, is a *Hydrate* of Carbon.
We cannot, however, produce artificially either
sugar, or any other organic compound, by directly combining their elements ; because we
cannot bring the elements together, precisely in
the requisite states and proportions.   Still, there
is no doubt, that if the elements could be so
brought together, the compound thence resulting,
would be the same as the natural compound.
For, as hereafter we shall endeavour to show,
the organic agent does not change the properties
of the elements ; but simply combines them in
modes which we cannot imitate.

Vinegar is another well known proximate principle, which does not only form crystallized com-

pounds readily, with many other bodies; but in
its most concentrated state, is itself also crystallized. Now, it is not less worthy of note than
in the case of sugar, that vinegar, altogether so
different from sugar in its properties, is generally considered to be precisely analogous in its
composition; that is to say, vinegar is a binary
compound of water and carbon; but the proportions of water and carbon are different from
those that form sugar. There is however, a
characteristic distinction between these two substances, inasmuch as vinegar can be formed
artificially; not indeed, any more than sugar,
by directly associating its elements: but, by the
process of fermentation, and by other means,
this acid may be formed from sugar, and from
the allied substances to be presently mentioned.
Yet we cannot work backwards, and by any
artificial process again form sugar from vinegar;
though the organic agent seems to possess this
power, as we shall have occasion to notice more
particularly hereafter.

We now proceed to consider the composition
of a totally different class of substances, which
under no circumstances, natural or artificial, ever
assume the crystallized form; and the structure
of which, in the common and strict sense of the
term, may be said to be *organized*. *Starch* is
a well known instance of these uncrystallizable
or organized substances.

The amylaceous or starchy principle is obtained in slightly modified states, from a great variety of vegetables, but principally from the seeds of the *Cerealia*. Even by the unassisted eye, starch is seen to be composed of minute particles; and when these particles are examined with a microscope, they are found to be granules more or less rounded, and without the least trace of crystallization. These granules are conceived to be moulded in the cellules of the texture by which they are formed; for it would appear that their state when first secreted and deposited in the cellules, is semifluid; and that the excess of water is subsequently removed. Raspail and Dumas have shown that each of these little grains is covered with a smooth integument, not affected by water at the common temperatures; within which integument is enclosed a substance rather more soluble. According to some chemists, this interior substance has an analogy with gum; but probably it is only a variety of amylaceous matter. Berzelius affirms that starch when burnt, leaves about ·23 per cent of residuum, consisting entirely of the phosphates. But when this residuum is abstracted and allowed for, the essential composition of starch is found to coincide very nearly with the essential composition of sugar; that is to say, starch is composed of water and carbon; and the proportions of their combination are very nearly

the same as in sugar. Here a question arises:
How does it happen that substances which appear
to resemble each other so closely in their composition, should yet differ so widely in their
sensible properties? This question we shall
soon consider. But in the mean time, we shall
make a few remarks on another principle of
organized bodies, still very different, in its sensible properties, from the three of which we
have spoken, but apparently of a similar constitution. This fourth principle is the *woody*
fibre, or *Lignin*, as it is termed by chemists.

The woody fibre, though assuming a great
variety of appearances in different plants, and
including very different incidental matters; has
nevertheless, in all those plants in which it has
yet been examined, been found to possess very
nearly the same essential composition ; or to
consist of equal weights of water and of carbon.
Such, at least, is the composition of woods, so
very different as the Box and the Willow, the
Oak and the Beech ; and these are the chief,
if not the whole, of the woods which, we believe,
have yet been analyzed. Hence, it is perhaps
not unreasonable to suppose, that every variety
of Lignin has a similar composition. All woods,
when burnt, leave a greater or less quantity of
incidental mineral residuum, in the shape of
ashes; the nature of which, as above observed,
differs exceedingly in different sorts of wood.

The following Table presents a view of the

composition of the four organic principles which we have considered in the preceding paragraphs. It is offered, not only as an example of the boundless subject of the Chemistry of Organization, but as an instance of the mode by which we conceive, that department of Chemistry may be best elucidated.

Substances Crystallizable.	Carbon.	Water.	Substances not Crystallizable.	Ca.bon.	Water.
SUGAR, from Starch	36·20	63·80	STARCH, Arrowroot in its ordinary state ..	36·4	63·6
from Honey	36·36	63·63	from Wheat, in its ordinary state ..	37·5	62·5
East India Moist..	40·88	59·12	ditto, ditto, dried at 212°.	42·8	57·2
Beet root and Maple	42·10	57·90			
English Refined ..	41·5 to 42·5	58·5 to 57·5			
Pure Sugar Candy.	42·85	57·15			
			LIGNIN, in its ordinary state of dryness	42·7	57·3
ACETIC ACID	47·05	52·95	from Willow, dried at 212°.	49·8	50·2
			from Box, ditto ..	50	50

A cursory inspection of the foregoing Table will evince to the reader, how nearly the general composition of sugar and of starch agree together; and that the agreement extends even to their several varieties. Vinegar, or acetic acid, has not, at present, any known representative, among other organic principles; though it is not improbable that several substances exist of conformable proportions. The composition of vinegar, or acetic acid, is intermediate to the composition of sugar, and of Lignin; while among crystallizable organic substances, there is no known compound analogous to Lignin. It may, at the same time, be remarked, that both starch and wood can, by different artificial processes, be converted either into sugar, or into vinegar. We can also convert wood into a sort of starch, as we may convert sugar into vinegar; but we are unable to reverse the process, and convert vinegar into sugar, or starch into wood; though these, and innumerable changes of a similar kind, are easily effected by organic agency.

We proceed now to consider briefly the question we have already stated,

2. *How does it happen that substances, so nearly allied in their composition, exhibit sensible properties so entirely different?*—This question, in all its bearings, is probably beyond our powers of investigation: at least the extent of the requi-

site knowledge we have yet attained, must be allowed to be exceedingly inadequate. The few observations which we have to offer regarding this question, may be comprised under the two following heads :—The *peculiarity of the composition* of organic substances ; and the *nature of the agents* by which these substances are produced.

The *composition* of organized bodies may be viewed as of two general kinds, viz. their composition, as depending simply upon differences among the proportions of their essential elements ; and their composition as depending upon differences among their incidental elements, the proportions of the essential elements being the same.* As instances of the first kind of composition, we may mention sugar and vinegar. Thus, sugar is composed of 42·85 per cent of carbon, and the rest water ; while the same ingredient, carbon, in the larger proportion of 47·05 per cent, with the residue water, constitutes vinegar, a powerful acid. Why, with such similarity of composition, the sensible properties of these two substances should be so unlike, we know not ; any more than we know why oxygen and hydrogen, when combined, form water ; or than

---

* Of course there is a third, and perhaps the most extensive class of bodies, in which both the essential and the incidental elements may be supposed to vary ; but partly from want of data, and partly to avoid too much complication, we shall not enter on the consideration of this class at present.

we know any ultimate chemical fact. However wonderful, therefore, the results of these slight differences of composition may, at the first view, appear; a little reflection will convince us, that in reality, they are not more wonderful than any other chemical phenomenon; and that they only form a particular variety of such phenomena. The same remarks are applicable, in part at least, to the striking differences exhibited by Sugar and Starch; the essential composition of which two substances, as we have before observed, is nearly the same; but the starch contains incidental bodies, from which the sugar is free. On the operation of these incidental bodies we shall offer a few conjectural remarks.

At the commencement of this chapter, we stated, that the incidental substances existing in organized bodies, have hitherto been considered as foreign; but that we could not subscribe to that opinion. We may now add, that the differences observed among bodies having the same essential composition, and which are at first view so mysterious; appear to us to be chiefly owing to their incidental ingredients. We cannot precisely understand the mode of operation of quantities so minute; but we can imagine them to be interposed among the constituent molecules: further, the molecules of these incidental matters, are probably in a state of strong self-repulsion. Such being the case, it is not unreasonable to expect, that these incidental matters may have the power of modifying the arrangement of the constituent molecules; and thus of altering the sensible properties of the substance produced by their combination.

We have stated our opinion, that the molecules of incidental matters in organic substances are in a state of self-repulsion. This opinion is founded principally, on the equal diffusion of these incidental molecules throughout the organic substances in which they exist; and on their consequent great distance from each other, which, perhaps, can hardly be otherwise explained. If these incidental matters were detached, or merely in a state of mixture with the constituent elements, as is implied in the notion of their being foreign; they would probably retain their self-attractive powers; and instead of being equally diffused among the constituent elements, they would be collected together into a mass or crystal; an arrangement never observed. For, though crystallized bodies are found, not unfrequently, within organized substances; yet these bodies are always extraneous, and do not form any part of the living structure; of which, the molecules under our consideration do actually appear to be integrants. In further corroboration of this opinion, may be adduced the beautiful experiments of Sir John Herschel, who has shown, that an enormous power, not less than 50,000 times the power of gravity, is instantaneously generated by the simple agency of common matters submitted to galvanic influence; as, for example, by the agency of mercury alloyed with a millionth part of its weight of Sodium. These facts, while they place beyond all doubt, the efficacy of minute quantities of matter, in producing the most extraordinary change of the polarities of larger quantities; at the same time appear to throw great light on many natural operations. Thus the subtle matters of contagion and miasmata; various medicinal substances, whose effects are most astonishing even in the smallest doses; the still more refined and recondite matters of heat and of light, with many others, all probably act on similar principles. At least, the results of the operation of these matters cannot be explained by their mere quantity; which, in the ordinary chemical acceptation of the term, is altogether incommensurate with the evident and striking changes, constantly arising in the processes of nature, from such agency.

The observations which have now been offered, are intended to apply to all the elementary substances, entering into the composition of a living organized being. For, no one element, when assimilated to a living body, appears to be in its natural state; or to be capable of exerting precisely those powers which it is known to exert, when acting in virtue of its original inorganic properties. In short, we may thus recapitulate what has been said : besides the essential molecules constituting the ground-work of a living organized being, and which probably exert on each other, to a certain extent, the ordinary chemical influences of matter; it would seem that there are, at the same time, diffused throughout the whole living mass, in exceedingly minute proportion, various other matters, the molecules of which appear to be in a high-state of self-repulsion. By these incidental matters, it would further seem, that the ordinary chemical properties of the essential elements of the organized living structure are variously modified; in particular, that, the essential elements are hindered from assuming a regularly crystallized form. Moreover, these incidental matters entering into the composition of a living body, apparently furnish to the organic agent new powers utterly beyond our comprehension ; which powers the organic agent has been endowed with the ability to control, and direct, in any manner, that from the exigencies of the living organized being, may become requisite.*

* In addition to what is stated in the text, we may remind the reader of what we have elsewhere alluded to, viz. that the organic agents have probably the power, within certain limits, of separating the molecules of bodies, considered at present as elementary, into more refined forms of matter (submolecules?).

The intimate nature of the *organic agent* or *agents*, or by whatever other name we may choose to designate the peculiar energies which exist in plants and in animals, and by which they are distinguished from inanimate matter; is now, and probably will ever remain, altogether unknown to us. But though we be thus ignorant of what these agents are ; we can not only comprehend with tolerable certainty, what they *are not ;* but we can also in some degree ascertain, what they are capable or incapable of effecting. As it is of the utmost consequence to obtain just views on these points, we shall consider them somewhat in detail.

When we were treating of inorganic elements and agencies, and of the laws which they appear mutually to obey, we found, that though their nature be obscure, and the investigation of them very difficult ; we were nevertheless enabled to adduce some, not altogether unplausible, conjectures on the modes, in which the elements combine to form regular crystals and the other conditions of inanimate matter. Now with this insight into the nature of inorganic operations, and with all the additional knowledge of every kind, we can command, let us attentively survey the most simple plant or animal ; let us observe the actions, the changes, the modifications of form and properties it continually exhibits ; and then let us seriously ask ourselves, whether every thing we know, will enable us to make, even an approach, toward an explanation of what we see. It is indeed true, that the plant or animal we examine is composed of charcoal and water, and of other ingredients with which we are equally familiar; that it is liable to be affected by Heat, Light, Electricity, and by other inorganic agents. But it is perfectly ascertained that these elements and agents, out of an organized body, *and left entirely to themselves*, never would or could unite, either in virtue of their own properties, or from accident, so as to form any plant or animal however insignificant. Are we not then compelled to infer, that within a plant or animal, there exists a principle, or agent, superior to those whose operations we witness in the inorganic world ; and which agent moreover possesses, under certain restraints, the power of controlling and directing the operations of these inferior agents? That this is a natural and a just inference, no one who calmly views all the circumstances will ever deny : and if the existence of one such agent be admitted, the admission of the existence of others can scarcely be withheld ; for the existence of one only, is quite inadequate to explain the infinite diversity among plants and animals. Thus, in the words of the excellent Paley, " there may be many such agents, and many ranks of them:" in other words, there may be an ascending gra-

P   F F

dation of these agents, from the vital agent in the comparatively simple plant, onward to that of the most complicated animal.

Such being the suggestions concerning organic agency, which arise from a general survey of organic operations; let us, with reference to the further bearing and tendency of these suggestions, enquire a little more minutely into the powers and modes of operation of organic agents.

3. *Of the Modes of Operation of Organic Agents.*—In the first place, with regard to what *cannot* be effected by organic agency, we may observe, that no organic agent has the power either of *creating* material elements, or of *changing* one such element into another. By element, it may be right to premise, is here meant, a principle which is not made up of others; and which, consequently, possesses an absolute and independent existence. Whether one, or more, such elements exist, it is not now our object to enquire. The astonishing discoveries of modern chemistry have shown, that many of those substances, formerly considered as elements, are, in fact, compounds; and as the science of chemistry is still progressive, it is probable, that with the enlargement of its boundaries, there will be a still further diminution of the number of those substances which are, as

yet, held to be simple. Admitting, however, for the sake of argument, that elementary principles do exist, of such immutable character as has been supposed: from the nature of organic beings, at least of all animals, it is impossible to conceive, that they possess the power either of creating or of altering these elementary principles. No organized being has an independent existence: and all animals derive their support from previous organization ; which might be otherwise, did they possess a creating power: nor can animals be nourished by any substances indiscriminately ; as they ought to be, were they possessed of a transmuting power. Yet, while it is thus denied that organized beings possess the power, either to create or to change, in the strict acceptation of these terms; it has been admitted to be exceedingly probable, that the organic agent is, within certain limits, qualified to compose and decompose many substances which are now viewed as elements ; and that the organic agent does thus apparently form and transmute these imagined elements. But to enter further, in this place, on the elucidation of these obscurities would be foreign to our present purpose.

The organic agent has not the power of combining elements in such a manner, that the properties of the resulting compound, shall *differ* from those of a compound, formed from the same elements similarly combined by any other agent.

The Deity has chosen to prescribe limits to his power, and to establish certain laws, to which He at all times rigidly adheres; and, again adopting the language of Paley, " when a particular purpose is to be effected, it is not by making a new law, nor by the suspension of the old ones, nor by making them wind, and bend, and yield to the occasion; but it is by the interposition of an apparatus corresponding with those laws, and suited to the exigency which results from them, that the purpose is at length attained." In the instance before us, the attainment of the particular purpose of organic life is effected, not by any departure from the great scheme, but by new and different combinations. To suppose, therefore, that the organic agent can, for example, combine oxygen and hydrogen, in exactly the same proportion, and in the same manner, in which they are combined, when they exist as water; and, from these elements so combined, can yet produce something different from water, is contrary to all reason, and would be, in truth, to accuse the Deity of subverting, and of acting in opposition to, his own laws. We have dwelt the more strongly on these points, because among physiologists a vague notion seems to have prevailed, that organic agents have the power, not only of changing the inherent and peculiar properties of bodies; but likewise, of causing the results of their combination to be altogether different from the results which are produced, under exactly similar circumstances, by inorganic agency. If however the arguments we have advanced be well founded, this notion must be erroneous; and its erroneous character will be rendered still more evident, by the observations, we shall, in the second place, offer, regarding the principles on which the operations within living organized bodies are really conducted.

The means by which organic agents accomplish the purposes for which they are designed, may be naturally divided into two kinds; those which are dependent on *peculiarity of composition and of structure;* and those by which *this peculiarity of composition and of structure is produced.*

Enquiry into the *first* of these means of action has already been in a great degree anticipated. A brief recital, therefore, is all that is here necessary. We have seen that organized substances are composed of the same elements, which exist abundantly throughout the world in the inorganized state; moreover, that these elements are subject to all the influences and agencies of inorganic nature. We have seen that organic agents are enabled to form certain proximate principles, by variously combining these elements; which proximate principles, even when in the condition of crystals, it is not possible to imitate artificially. We have, at the same time, seen that these proximate principles, though they may have a natural tendency to crystallize, are, as they usually exist in living bodies, prevented from crystallization, by having minute quantities of various other elements diffused throughout their mass ; the molecules of which diffused elements are in some unknown state of activity ; such perhaps as cannot naturally exist in the universe, except when conjoined with organization. Finally, we have inferred, that the differences and peculiarities of these minute additional matters, are probably adequate for explaining the differences and peculiarities, of the sensible and chemical properties of the substances which are formed by organization. Having thus pointed out the general differences of composition existing among organized bodies ; it remains to state, that such differences of composition almost invariably indicate differences of structure. For though similarity of composition does not necessarily imply similarity of structure ; yet similarity of structure, perhaps, without exception, indicates similarity, or, at least, analogy of composition ; and, consequently, *similarity of action.* Thus the woody fibre of plants is always formed of the principle termed *Lignin*, and never of resin, or of albumen. The relation of structure to chemical composition is not less striking in the muscular fibres of animals, and indeed in all organic compounds of a definite character; the essential composition of such substances, though exhibiting endless minor diversities, being, nevertheless, in all instances, precisely the same.

The means by which that *peculiarity of composition and of structure is produced*, which is so remarkable in all organic substances, are, like the results themselves, quite peculiar; and bear little or no resemblance to any artificial process of chemistry. For example, we have not, in artificial chemistry, any control over individual molecules; but are obliged to direct our operations on a mass, formed of a large collection of molecules: The organic agent, on the contrary, having an apparatus of extreme minuteness, is enabled to operate on each individual molecule separately; and thus, according to the object designed, to exclude some molecules, and to bring others into contact. In these processes, it may be conceived, that the molecules thus appropriately brought together, and, at the same time, guarded from extraneous influence by the organic agent, are in virtue of their own proper affinities, sufficiently disposed to unite, without requiring that any new properties should be communicated to them. Hence the organic agent, in its simplest state, may be viewed as a power so controlling certain inorganic matters, as to form them into an apparatus, by which it arranges and organizes other matters, and thus

effects its ulterior purposes. Where the operations of this simple organic agent terminate,
those of another and more effective organic
agent may be supposed to begin; which, by
carrying the general process of organization a
step further, adapts the organized material for
the operations of a third and yet higher agent.
Thus, each new agent may be supposed to possess more or less control over all the agents below
itself, and to have the power of appropriating
their services; till at length, at the top of the
scale, we reach the perfection of organized
existence. The excellent Paley sanctions this
view of organic operations, and continues in the
following words : " We do not advance this as a
doctrine either of philosophy or of religion ; but
we say that the subject may safely be represented under this view; because the Deity,
acting himself by general laws, will have the
same consequences upon our reasoning, as if he
had prescribed these laws to another."

This view of the successive creation of organic
agents, which harmonizes not only with the
phenomena of Geology, but with the differences
which are observable among plants and animals,
and with the developement of the more perfect
species ; is directly opposed to the notion of spontaneous developement maintained by some distinguished French philosophers ; as well as to the
opinion that life is the *result* of organization.

The laws of nature, as we have shown, are in all cases most rigidly adhered to by the Deity. These laws, therefore, are unalterably stable, within the limits, which have been assigned to them.   Now, from what we know of the laws of nature, or of the properties of the elements of matter, or of the agents by which they are moved; it is, as we have already stated, impossible to conceive that carbon, water, and electricity, of their own accord, and from any inherent influence, can so unite as to form the humblest plant or animal; much less, so as to secure its perpetual existence by reproduction.   For similar reasons it is equally impossible to conceive, that there can ever be such a spontaneous arrangement or combination of *inferior* organic agents, as to form a *superior* agent: and when a new and specific agent is required, a new and specific act of creation must be performed by the Great Architect of the universe.   We consider it, therefore, impossible that by any accidental concurrence of circumstances, a dog can, in the progress of time, be gradually converted into an ape, or an ape into a man; and moreover, we not only think such an hypothesis directly at variance with the whole tenor of the laws of nature, but quite absurd.

Nearly similar remarks apply to the opinion that the living principle is the *result* of organization. The living principle is *not* the *result* of organization, but the *cause* of organization. In accounting for the phenomena of life, it is absolutely necessary to assume the existence of some agency different from, and superior to, that which operates among inorganic matters. Now since, as we have seen, no inferior agencies can be supposed so to combine as to form a superior agency; does it not accord better with our reason, as well as with our experience, to assume at once a new creation of the higher principle?

In regarding the nature and composition of organized bodies, the first circumstance which arrests our attention, is the wonderful adaptation of the elements and agents of inorganic, and of organic nature, to each other. For example, had not carbon, and azote, and water, been formed with the properties which they now possess, organic agents, as we know them, would have existed in vain; and without organic agents, the properties of these elements would equally have been useless. And how truly wonderful, and utterly beyond our comprehension, are the properties and adaptations displayed in the processes of organization! To enable ourselves to form some conception of these processes, by bringing to a level with our understanding, those things which they accomplish; let us propose to

ourselves the question,—What ought to be the
inherent properties and the constitution of an
elementary principle, which should not only be
capable of being formed into the hardest and
the softest bodies in nature; but which should
also be capable of entering as an essential ingredient into substances so very unlike, as sugar,
vinegar, wood, oil, albumen, and many others, in
all their countless forms and varieties? Do we
not feel all our fancied knowledge annihilated
by such a question? Nay, what is more, even
when the question is answered for us; and when,
with the utmost care, and to the furthest extent
of our ability, we have studied all the chemical
properties of Carbon—the substance by which
the conditions of the question are fulfilled; how
totally unable are we to explain these properties,
or even to trace them through their simplest
modifications? Why, for instance, is the diamond
capable of assuming the form of charcoal; or
why is charcoal capable of assuming the form
of the diamond? And how are these properties
modified, and altered, in all the numerous states
of combination into which we know carbon
enters? On what property or quality, not possessed by other elements, do all those astonishing capabilities of change depend, which are
inherent in this element carbon? And why
has carbon been chosen for forming organized
beings, in preference to silex, or iron, or any

other element ?* To us all these things are absolutely unknown ; but what a conception do they give, of that inscrutable agency by which the elements are governed ; of the powers of that Almighty Mind who is conversant with them all—by whom they were first designed, and by whom they have all been created! How infinitely must His knowledge surpass whatever we can imagine : how far is His power beyond our utmost calculation !

On the other hand, if the properties of the ele-

---

* Since there is nothing peculiar in the elements of which organized beings are composed, and no reason can be assigned why carbon and other elements have been chosen for their formation, we are compelled to ascribe the choice of these materials to the will of the Great Creator. But as He never acts without a purpose, we cannot doubt that these elements have been selected for some specific design ; which design has probably been, that the fabric of the beings dwelling on this earth, might be adapted to its general position in the Solar system. When we consider that the same heat, and the same light are diffused by the same central sun ; that the whole system obeys the same laws ; and that the different planets influence, and are influenced by each other ; we are warranted in believing that the planets are essentially composed of the same elementary principles. But admitting that the heat and light of the sun are distributed according to the laws which they seem universally to obey ; the heat in Mercury, close to the sun, and the cold in Saturn, at the other extreme, must be alike so intense, that organized beings, such as inhabit this earth, could not exist for a moment. In the different planets, therefore, may not the living principle be attached to different elements, more or less fixed or volatile, as the distance of the planet from the sun may require ?

ments of matter be wonderful ; yet more wonderful are those agents within organized bodies, by which they are directed. With the intimate nature indeed of these agents, we have not the remotest acquaintance, nor, probably, ever shall have. But, as has been already stated, we can trace, to a certain extent, the laws of action which organic agents obey : we observe the unvarying adaptation of these agents to the properties of carbon, azote, and water, on which they chiefly act ; their power, within certain limits, of guiding and controlling inorganic agents ; and more than all, that mysterious periodic developement and decay, which every organized being undergoes. These facts which continually present themselves to our notice, are totally inexplicable according to those laws by which inorganic bodies are governed ; and are referrible only, to an order of laws, which have not been revealed.

Lastly, we cannot close this chapter, without pointing out to the reader a very remarkable contrast, in the two classes of objects which have engaged our attention. The number and diversity of organic agents appear to be endless; in the creation, therefore, of these agents, the Great Author of Nature has chosen to manifest his attribute of infinity. But in the creation of the material elements which compose the frame of organized beings, He has adopted a plan directly opposite. Instead of different principles; the same carbon, the same azote, the same water, enter into every living being, from the lowest plant upward to man. Amidst the wonders of creation, it is perhaps difficult to say what is most wonderful; but we have often thought, that the Deity has displayed a greater stretch of power, in accommodating to such an extraordinary variety of changes, a material so unpromising and so refractory as charcoal, and in finally uniting it with the human mind; than was requisite for the creation of the human mind itself. But, to Him, all things are alike easy of accomplishment; and He, doubtless, has willed these and other proofs of His omnipotence, in order to convince us of this truth, —that the Creator of the mind, could alone have created the matter with which the mind is associated!

CHAPTER II.

OF THE MODES OF NUTRITION ; COMPRISING A BRIEF
DESCRIPTION OF THE ALIMENTARY APPARATUS ;
AND OF ALIMENTARY SUBSTANCES, IN PLANTS AND
IN ANIMALS.

THE subsistence of all organized beings is derived from sources external to themselves ; and
the sources of their aliment as well as the modes
in which these aliments are applied, exhibit an
almost endless variety. As might be expected,
the widest differences, both in the nature of
the alimentary substances, and in the manner
of their introduction, are between plants and
animals. We shall, therefore, consider the subject of nutrition under these two heads.

SECTION I.

*Of the Modes of the Nutrition of Plants; and of
the Nature of those Matters by which their
Nutrition is effected.*

A MINUTE investigation of the anatomy and the
physiology of plants, would be quite foreign to

the object of this treatise. At the same time, it is necessary that the reader should have some insight into these departments of knowledge; in order that he may be enabled to understand the collateral researches, which it is our duty to illustrate.

" If we reflect upon the phenomena of vegetation," says Professor Lindley, " our minds can scarcely fail to be deeply impressed with admiration at the perfect simplicity, and, at the same time, faultless skill, with which all the machinery is contrived, upon which vegetable life depends. A few forms of tissue interwoven horizontally and perpendicularly constitute a *stem;* the developement, by the first shoot that the seed produces, of buds which grow upon the same plan as the first shoot itself, and a constant succession of the same phenomenon, causes an increase in the length and breadth of the plant ; an expansion of the bark into a *leaf,* within which ramify veins proceeding from the seat of nutritive matter in the new shoot, the provision of air passages in its substance, and of evaporating pores on its surface, enables the crude fluid sent from the roots to be elaborated and digested until it becomes the *peculiar secretion* of the species: the contraction of the branch and its leaves forms a *flower;* the disintegration of the internal tissue of a *petal* forms an *anther;* the folding inwards of a leaf is sufficient to constitute a *pistillum;* and finally, the gorging of the pistillum with fluid which it cannot part with, causes the production of a *fruit.*"*

The "crude fluid sent up from the roots" of plants, or their *sap*, as it is termed, is found to consist of water, mucilage, and sugar; with some minute portions of other matters, generally saline. Though, under certain circumstances, moisture is absorbed by the leaves of all plants, yet there can be no doubt that a great part of their nourishment enters by their roots; not, however, by the whole root indiscriminately: the nourishment of plants is taken up chiefly by the minute fibrous parts termed *spongioles*. Hence, these minute fibrous parts are of the utmost importance in the vegetable economy, and ought to be carefully preserved in transplantation; otherwise the plant will certainly perish. In some instances, roots appear to be intended to act as reservoirs of nourishment for the support of the plants of the succeeding year, on their first developement. There are such roots in the *Orchis* and *Dahlia* tribes, and in others. Of late it seems to have been satisfactorily established, that the roots of all plants, besides imbibing nourishment, perform also an excretory office; and that in the soil in which plants grow, there are deposited by the roots, certain matters of an

* Introduction to Botany, p. 216.

P.   G G

excrementitious nature; injurious to the plants from which they have been separated; and which matters, therefore, cannot be absorbed again, till they have undergone decomposition. Such excreted matters have been adduced as the reason, why a soil becomes, sooner or later, so much deteriorated by any one species of plant, that it will not support other individuals of the same species : whence the necessity of a rotation of crops.

The principal ingredient in the sap of plants, as we have already stated, is water. The quantity of sap in some plants, is almost incredible ; and not less so, is the force with which, on the approach of warm weather in our climates, and at the commencement of the rainy season within the tropics, that sap is determined upwards. The general composition of the sap varies considerably in different parts of the same plant. For instance, sap taken from the roots is little more than water; while the quantity of saccharine and other matters contained in the sap, increases in its progress along the stem to the higher parts of the plant. When the sap begins to rise, the leaves at the same time begin to be developed. From the leaves principally the watery portions of the sap are evaporated ; and the evaporation is copious and unceasing. The more solid matters thus remain dissolved in a less proportion of water ; and after undergoing further changes, in the leaves chiefly, as is supposed ; these matters are returned, along with the remaining water, to be deposited in other parts of the plant, for its future uses. It seems now to be generally admitted, that one part of the food of plants is the matter extracted from the soil ; and that this matter is taken up with the watery portion of the sap above mentioned. It seems also to be admitted, that carbonic acid gas is in some way indispensable to vegetation ; " for it has been ascertained, that feed plants as you will, they will neither grow nor live, whether you offer them oxygen, hydrogen, azote, or any other gaseous or fluid principle, unless carbonic acid is present." Like the other nutritious matters, this carbonic acid is partly taken up by the roots ; but under certain circumstances, it is also absorbed from the air, by the leaves. The circumstances under which this absorption, or rather decomposition, of carbonic acid, by the leaves, takes place, are most curious and important. They are understood to be as follows :

During the day, and particularly during sunshine, the leaves of plants have the power of abstracting the carbonic acid from the atmosphere. The carbon of the acid, and perhaps also a little of its oxygen, combine with the plant ; while the greater part of the oxygen remains, and is diffused through the atmosphere in a gaseous state.

During the night, on the contrary, or in the shade, plants, in general, convert a portion of the oxygen of the atmosphere into carbonic acid; but the quantity of oxygen thus converted into carbonic acid, is less than the quantity of oxygen separated from the carbonic acid, which plants decompose, under the influence of the solar light. At the same time with this formation of carbonic acid during the night, plants are said also to absorb from the atmosphere a certain portion of oxygen; to replace that oxygen which had been given off, during exposure to sunshine, on the preceding day. Plants absorb carbon as long as they are exposed to the light: during the season, therefore, when the day is long and the night is short, plants give off much less carbon than they absorb. This excess of the absorption of carbon, is probably one reason why in the Polar latitudes, the progress of vegetation is so rapid. By a beautiful provision of nature; in the course of the short summer of a few weeks, but of unvarying light; plants, in these latitudes, go through all the changes which in hotter climates require many months.

These phenomena of gaseous absorption and secretion in the leaves of plants, seem to be produced by a portion of the leaf peculiarly organized, and situated immediately under its external covering or epidermis. Professor Burnet has lately explained these phenomena, by referring them to the respiration and digestion of plants. The process of respiration in plants, is supposed to be continual; and to be accompanied, as in animals, by the formation and emission of carbonic acid gas. The process of digestion in plants, on the contrary, takes place only during their exposure to the light of the sun: their digestive process consists in the decomposition of the carbonic acid gas of the atmosphere, and the absorption of the carbon from the acid which is thus decomposed. Hence a plant, under the influence of sunshine, purifies the air by digesting the carbonic acid, and appropriating the carbon; while in the dark, the digestive process of plants ceases; but they continue to respire without intermission; and carbonic acid gas is thus accumulated in the surrounding atmosphere.

With respect to the " peculiar principles of plants," these are as numerous as the individual plants themselves; so that to attempt any detailed account of them here, would be quite impracticable. Generally speaking, the peculiar principles obtained from plants may be divided into three great classes :—those vegetable principles arising from the combination of hydrogen and oxygen in the same proportions which constitute water; as in the division of *saccharine* bodies, described in a former chapter :—those principles in which hydrogen, or rather carbon and hydrogen, predominate; which generally have more or less of an *oily* character;—and those principles in which oxygen predominates; which have usually an *acid* character. Besides these three great classes of vegetable principles; there are some which contain azote, and perhaps other elements. Many vegetable principles, also, exhibit weak alkaline powers : such are the peculiar principles of Opium and other Narcotics; also of Cinchona; and a variety of others, chiefly employed in Medicine.

Section II.

Of the Modes of Nutrition in Animals; and of the Alimentary Substances by which they are nourished.

To beings, like animals, endowed with locomotive powers, the absorption of their nourishment from without, would have been exceedingly inconvenient. Animals have, therefore, been furnished with an additional receptacle and apparatus subservient to nutrition; into which, as inclination or circumstances may prompt them, their food is conveyed at intervals; and from which, after having undergone certain changes, the food is absorbed, and distributed over their system, as the exigencies of that system may require. Hence the distinction between

plants and animals;—plants absorb their nourishment by *external,* animals by *internal,* roots
or *spongioles.* We need scarcely remark, that
the stomach and alimentary canal, with their
appendages, are the internal apparatus to which
we allude ; and that this internal apparatus constitutes a marked difference between plants and
animals.

1. *Of the Organs of Digestion in Animals,*—
Among the different tribes of animals, there is
an almost endless diversity in the formation of
the alimentary organs ; and as these organs vary,
not only in their own formation, but also with
respect to the auxiliary apparatus, and appendages of every kind, connected with them ; any
detailed account of the alimentary system would
at present be quite uncalled for. In general,
the alimentary canal of the higher classes of
animals, consists of a tube of greater or less
elongation; expanded in some parts of its length;
terminated at one extremity by a mouth, into
which the food is received ; and at the other,
by a provision for the removal of excrementitious
matters. In some of the less perfect animals,
the alimentary canal has only one aperture:
in such animals, of course, instead of a canal,
there is a kind of sac. In a very few other
animals, the alimentary cavity has numerous
apertures. In all instances, however, and what-

ever may be the nature of the alimentary matters; these matters, after having been retained for some time in the organs appropriated to nutrition, are reduced, more or less, to a fluid state—are DIGESTED, in the common sense of the term, and are converted into what is denominated *chyme.* The more nutritious parts of the fluid chyme, or the *chyle* as they are denominated, are then absorbed, and distributed through the system for the reparation of the animal; while the insoluble and other matters, are separated as excrementitious.

We have already alluded to the endless diversity observable in the form and arrangements of the alimentary canal in the different kinds of animals. A few of the most remarkable of these diversities among the more perfect animals will be noticed, in the following outline of the alimentary canal as existing in the human body.

Of the Mouth and its Appendages.—" In no apparatus put together by art," says Paley, " do I know such multifarious uses so aptly contrived as in the natural organization of the human mouth." " In this small cavity we have teeth of different shape,—first for cutting, secondly for grinding; muscles most artificially disposed for carrying on the compound motion of the lower jaw, half lateral and half vertical, by which the mill is worked; fountains of saliva springing up in different parts of the cavity for the moistening

of the food, while the mastication is going on;
glands to feed the fountains; a muscular construction of a very peculiar kind in the back
part of the cavity, for the guiding of the prepared
aliment into its passage towards the stomach, and
in many cases for carrying it along that passage."
"In the meantime, and within the same cavity,
is going on another business altogether different
from what is here described—that of respiration
and of speech. In addition, therefore, to all that
has been mentioned, we have a passage opened
from this cavity to the lungs, for the admission
of air, exclusively of every other substance; we
have muscles, some in the larynx, and without
number in the tongue, for the purpose of mo-
dulating that air in its passage, with a variety, a
compass, and a precision of which no other mu-
sical instrument is capable. And lastly, we have
a specific contrivance for dividing the pneumatic
part from the mechanical, and for preventing one
set of actions interfering with the other." "The
mouth, with all these intentions to serve, is a
single cavity; is one machine, with its parts
neither crowded nor confined, and each unem-
barrassed by the rest."* Such is Paley's gra-
phic description of the human mouth and its
appendages: we have quoted it at length, that
it may serve as a text for illustration.

* Natural Theology, chap. ix.

Man has been observed to differ more from other animals in the form of his *lower jaw*, than in the form of any other bone of his body. This difference consists chiefly in the prominence of the chin ; that peculiar characteristic of the human countenance, which distinguishes more or less every race of mankind, and is found in no other animal whatever. There is likewise a striking difference, among the various tribes of animals, in the mode of articulation of the lower jaw ; which in all cases is singularly adapted to the nature of the food of the animal. Thus, in the carnivorous tribes, the articulation is so arranged that the jaw can move only up and down ; and is almost entirely incapable of that lateral movement, which is essential to genuine mastication. Hence such animals cut and tear their food, and swallow it in large pieces. But those animals that live on vegetables, in addition to the vertical motion of their lower jaw, have the power of moving it backwards and forwards, or to either side, so as to produce a grinding effect, admirably fitted for triturating the vegetable matters on which they subsist.

The *teeth* next claim our attention, as being not less suited to the habits of the animal, than the form of the jaw in which they are set. Teeth are divided by naturalists into three orders :—The *Incisores*, or *cutting teeth*, placed in the front part of the mouth ; the *Cuspidati*, *canine,* or *corner teeth,* usually placed near the angles of the jaw ; the *Molares, grinding,* or *lateral* teeth, which always occupy the sides and back part of the jaw. In man, and in those animals which most nearly resemble him in their structure, teeth exist of all the above varieties of form. But many species want one or other of these varieties ; while the teeth they possess, are of a form and size very unlike the same teeth in man. Thus, in animals which live chiefly on the harder vegetable substances, and which, from their peculiar mode of feeding, have been termed *gnawing* animals; the *incisor* teeth are the most remarkably developed ; as these teeth are the best adapted, and indeed are the most necessary, to their habits. In *carnivorous* animals, on the other hand, the *canine* teeth are of chief importance; as enabling these animals to seize and hold their prey : in such animals, accordingly, the canine teeth are the most perfectly formed. Lastly, in the animals that feed on grass, and other *herbaceous* substances, and whose aliments require long and complete mastication, the *Molares,* or grinding teeth, attain the greatest enlargement ; and in many of these animals, the incisor and the canine teeth are entirely wanting. Besides the adaptation of the form; the enamel or harder cutting portion of the teeth, is distributed over and throughout their texture, according to their intended uses, in a manner that is truly extraordinary. The description of the arrangement of the enamel, as well indeed as a minute account of the teeth themselves, belong, however, to the physiologist, on whose province we shall not further intrude. But it is impossible to take even the most superficial view of the teeth of animals, without being struck with the admirable design and fitness they display, throughout their whole fabrication.

The next auxiliary appendages of the mouth are the *glands* that secrete the saliva; in which we observe the same beautiful arrangement as in the form and structure of the teeth. In man, though the apparatus for the secretion of the saliva, is by no means of large size; yet the quantity of fluid which the salivary glands are capable of secreting, and do secrete during mastication, is very considerable; often amounting, it is said, to half a pint or more. This fluid, in its perfectly healthy state, is neither acid nor alkaline, or alkaline only in a slight degree; but occasionally it assumes an acid character. Besides the great utility of the saliva in moistening the food, we cannot doubt that it assists, and is even necessary to the full completion of, the succeeding digestive process. By a beautiful arrangement, animals which do not masticate their food, as the carnivorous tribes, have very small salivary glands; while in animals whose food requires long mastication, as in ruminating animals—the cow and the sheep, for example, the salivary glands are very large.

The passage by which the masticated food is conveyed from the mouth to the stomach, is termed the *œsophagus*. Like the whole frame, the œsophagus is admirably adapted for its office; and in different animals, varies in size and structure, according to their habits. These differences, however, scarcely concern us at present, and we pass on to that important organ —the Stomach.

The human *stomach* is a membranous bag, of a shape rather difficult to be described, so as to convey a clear notion of it to the reader. If we imagine two cones united at their bases, and the figure thus produced to be bent into a semi-circular form; some idea may be obtained of the outline of the stomach in the human species. In respect to its size; the human stomach varies: but in the adult, its capacity is usually such as to contain about two or three pints. The stomach is situated immediately under the diaphragm; but the precise place of the organ differs somewhat with its state of repletion. The general position of the stomach is transverse, or horizontal, supposing the body to be upright : the left orifice, or *cardia*, which communicates with the œsophagus, being somewhat higher than the right orifice, the *pylorus*, through which the food is transmitted to the further portion of the alimentary canal. The upper space between the two orifices is usually termed the small curvature; the lower space, the great curvature, of the stomach. Numerous glands occupy the internal surface of the stomach, particularly near its pyloric orifice. By these glands a fluid is secreted of the highest importance in the· digestive functions. On the nature of that fluid we shall enlarge hereafter.

Such is the stomach of man; but the form and the magnitude of the organ vary almost infinitely in different animals, according to the nature of their food, and other circumstances. We can, at present, notice only two or three of the most remarkable diversities. The stomach of most carnivorous animals, bears a resemblance to the stomach of man. There is also a resemblance, at least externally, in certain herbivorous animals; as in the horse, the rabbit, and others. The internal arrangements, however, are different; thus, in the animals above mentioned, the left or cardiac half of the stomach is lined with cuticle; while the other half, towards the pylorus, has the usual villous and secreting surface. Hence, these two portions of the stomach perform very different offices, and generally contain food in very different states of reduction. But the most complicated and artificial arrangements, both with respect to the structure of the parts, and the lining membranes, are found in the well-known four stomachs of the animals that ruminate and have divided hoofs; as the cow and the sheep. We shall endeavour to give a general description of these four stomachs. The first stomach is denominated the *Paunch*, and in the adult animal is by far the largest. The second stomach follows, and may be regarded as a globular appendage to the paunch; from which it is distinguished, principally, by the regular and beautiful distribution of its internal membrane into polygonal cells. The third stomach is the smallest of the four; and is the most remarkable in its structure : its capacity is much diminished by numerous and broad duplicatures of its internal membrane, which are placed lengthwise, and vary in breadth in a regular order. The fourth stomach is next in size to the paunch, and is lined with a villous membrane, similar to the villous membrane of the human stomach, which this fourth stomach may be supposed to represent; the three preceding stomachs having been evidently intended to prepare the refractory food of the animal for the true digestive process, undergone in this last stomach. Every one is acquainted with the fact that animals furnished with the gastric arrangements above described, *ruminate ;* that is to say, have the faculty of masticating a second time, and at their leisure, that food which had been hastily swallowed, and deposited in their first stomach. The contrivance by which rumination is effected is very beautiful; and is connected with the peculiar arrangement already mentioned of the four stomachs, with respect to the œsophagus: but, as it would not be easy to give, in few words, more than a general outline of the stomachs of ruminating animals; for a particular description of them, we must refer the reader to anatomical works. The only other modification of the stomach which we shall notice, is that which exists in some birds; as for example, in the common fowl. The common domestic fowl, as well as many similar birds, has a sort of preliminary stomach, termed the *crop*, formed by an expansion of the œsophagus. In the crop, the hard seeds, and other compact substances which birds devour, are macerated and softened, and perhaps undergo further changes, before they enter the proper stomach, to be next considered. The proper stomach, or *gizzard*, of birds, is a hollow muscle of great strength, lined with a firm and thick epidermis, disposed in rugæ, and admirably adapted for triturating the hard matters which constitute their food. The small stones, these birds constantly swallow, seem also to promote this triuration.

We have given the above short sketch of the structure of the stomachs of animals, not only that we might impart to the general reader a faint conception of the extraordinary design manifested in that structure; but to enable us to show the object of diversity of structure, when we come to speak of the function of digestion a little more in detail.

After the stomach we proceed to the consideration of the *Intestinal Canal.* In man, and in the more perfect animals, this canal assumes two well marked forms; usually termed, from their relative size, the small and the large intestines. In most animals resembling man, the small intestines are the longest, and their internal surface is villous. The coats of the large intestines are thicker, and the membrane with which they are lined is very rarely villous. The first portion of the small intestines, from its supposed length in man, termed the *duodenum*, or twelve-inch intestine, begins from the pyloric orifice of the stomach; and, in many animals has a course not easy to be described, so as to be intelligible to the general reader. The duodenum terminates in the second portion of the small intestines, called the *jejunum*, from its being usually empty. The duodenum is secured in its position by various attachments: in this fixedness, the duodenum differs from the stomach and other parts of the intestinal canal, which are comparatively loose and floating. The unaltered position of the duodenum appears to serve many wise purposes, on which we cannot dwell here; but one purpose probably is, to ensure the easy and regular passage of the bile and the pancreatic fluids into this part of the canal. As the organs producing these important fluids are fixed, the conducting tubes necessarily require also to be connected with a fixed organ; otherwise the passage of the fluids from the secreting organs to the intestine, would be constantly liable to interruption. The duodenum is very highly organized; and its functions are probably not less important than even those of the stomach. The remainder of the small intestines is divided into the *jejunum* already mentioned, and the *ilium;* but the precise place where one ends, and the other begins, is scarcely definable; nor are the differences of structure between the two so obvious, as to require to be noticed in this place.

The large intestines exceed the small intestines in diameter, but are considerably shorter: their form and structure are also different. The first division of this portion of the alimentary canal is termed the *cæcum;* and, in man at least, may be considered as little more than the head or commencement of the next division of the large intestines, termed the *colon.* The colon is of much greater diameter than any other part of the intestinal canal, and constitutes almost the entire length of the large intestines. The colon begins low down on the right side of the abdomen, then ascending to the level of the stomach, passes across to the left side, immediately below the

stomach. On the left side, the colon descends
again, and at the same time, forms what is called
the *sigmoid* flexure. The colon and the alimentary canal at length terminate in what is named
the *rectum*. The texture of the colon is much
thicker than that of any other portion of the
canal. Its organization also is peculiar; and,
like the whole arrangement, wonderfully adapted
for the purposes which this portion of the canal
is supposed to fulfil in the animal economy.

Such is a short account of the alimentary
canal in man. We shall now state some of the
more remarkable diversities that are observed
in the lower animals.

One of the most striking circumstances relative to the alimentary canal in animals, is its
various lengths in the different classes. The
length of the alimentary canal in man, and in
other omnivorous animals, is intermediate to that
of carnivorous animals on the one hand, and of
herbivorous animals on the other. In man, the
whole length of the canal is about six or seven
times the length of his body; while in carnivorous animals it is only from about three to
five times the length of their bodies; and in
graminivorous animals, as in the sheep, the
length of the canal is twenty-seven times that
length. In other herbivorous animals, the length
of the canal varies from twelve or sixteen times
the length of their bodies. In most birds the
alimentary canal is much shorter than in quad-

rupeds ; the length in general, being between twice and five times that of their bodies : while in many reptiles and fish, the length of the canal scarcely exceeds that of the body : in some fish it is even less; as for example, in the shark. There are animals, vegetable feeders, the length of whose alimentary canal is not so great, as in the instance above stated ; the deficiency of length being apparently made up in breadth. Thus, in the horse, the stomach is simple, and not much developed, when compared with the size of the animal ; nor are the intestines very remarkable for their length ; but the cœcum and the large intestines are enormously expanded in diameter. The cœcum of the horse seems to perform many of the offices of a second stomach, and is of fully equal capacity. There are in animals, many other beautiful arrangements of the digestive organs, which we shall pass without further notice ; as our desire is to inform the reader, merely of the general connection and adaptation, which exist between the structure of animals, and the food on which they live. It remains to conclude this outline of the digestive organs, with a few remarks on those almost invariable accompaniments of the alimentary canal,—the *liver*, the *pancreas*, and the *spleen*.

The liver is the largest glandular apparatus in the body ; and one of its important offices is to secrete the *Bile;* which secretion, as before observed, enters the intestines, near the commencement of the duodenum. The general situation of the human liver, is in the upper part of the abdomen, under the ribs on the right side ; from whence it extends more or less to the region of the stomach, and in some instances, even to the left side. The appearance and form of the liver, are too well known to require description here ; while to those who are unacquainted with these particulars, they cannot be adequately made known by words. In man, and the greater number of animals, the bile is collected in a small bag, termed from its office the *gall-bladder*. The animals wanting a gall-bladder are chiefly vegetable feeders ; as the horse and the goat among quadrupeds, the pigeon and the parrot among birds. On the contrary, most amphibia have a gall-bladder ; but it exists in few animals lower in the zoological scale. The liver assumes a variety of forms in different animals. In many, and particularly in carnivorous, animals, the liver is more divided than in man : while in ruminating animals, also in the horse, the hog, and others, its divisions are not more numerous than in man. The liver of birds consists of two lobes of equal size.

The *pancreas*, or *sweetbread*, is a large gland, which, in the human body, lies across the upper and back part of the abdomen, behind the stomach ; and between the liver and the spleen. The pancreas is composed of numerous small glands, whose ducts unite and form the pan-

creatic duct. In man the pancreatic duct joins
the gall duct, at its entrance into the duodenum;
and thus the peculiar secretion of the pancreas is
poured into that intestine, commingled with the
bile. In animals, the pancreas, like the liver,
is much varied in its form ; and its duct, instead
of entering with the biliary duct, often joins the
intestinal canal separately ; as in the hare and
others. In fishes the pancreas is wanting ; but
what are termed the cœcal appendages, are
supposed to have a similar office. The nature
of the pancreatic fluid will be considered presently.

The human *spleen* is situated in the upper and
left side of the abdomen. Its shape is oblong,
and its colour a deep mulberry ; more nearly resembling the colour of the liver than of any other
organ. The spleen has no excretory duct; and
its use is very little understood. Among the less
perfect animals, the spleen is much smaller than
in those whose structure resembles that of man :
and where there is more than one stomach, the
spleen is always attached to the first. The situation also of the spleen varies in the less perfect
animals: thus the spleen of the frog is fixed in
the mesentery.

We proceed to notice, very briefly, the peculiar circulation of the blood in the abdominal
viscera ; together with the character and agency
of that portion of the nervous system, which

is connected with the digestive and assimilating functions of animals.

In the general circulation of the blood through an animal body; a large tube or artery, communicating with the heart, is gradually subdivided as it is prolonged from that organ, till its subdivisions finally become imperceptable. The arteries, in this state of minute subdivision, assume the character of veins. The veins, in their progress, undergo a change, the reverse of the change undergone by the arteries: they unite gradually, and, at length, form one or two principal tubes, which proceed to the side of the heart opposite to where the artery originates. Such is the circulation of the blood through the body generally : the circulation through the lungs is merely a repetition of the same arrangement. Throughout the body, therefore, the general motion of the blood in arteries is from greater to smaller tubes ; while in the veins, the motion is from smaller to greater tubes. By a beautiful provision, the veins are also furnished with valves; which most effectually prevent the regurgitation of the blood : without such valves, the blood could scarcely flow in a regular stream. We have introduced these remarks, with the view of stating, that the circulation of the blood through the organs of digestion, presents a remarkable exception to the general circulation of the body. The venous blood, from the organs

of digestion, is submitted to a preliminary arte-
rializing process in the liver; before it is remin-
gled with the venous blood from the rest of the
body. That is to say, the veins from the organs
of digestion, unite into one large tube termed
the *vena portæ;* which tube, entering the liver,
is there again subdivided, in the same manner
as an artery. These ultimate subdivisions of
the *vena portæ,* together with the similar subdivisions of the proper artery of the liver, coalesce; and from the blood thus mixed, the bile
is separated. The coalesced blood vessels assuming the character of veins, then gradually
unite, and at length form two or three large
tubes, which empty themselves into the general
veins going to the heart; while the hepatic ducts,
uniting in like manner, convey the bile to the
gall-bladder. Such are the principal facts connected with the circulation of the blood in the
abdominal viscera; and with the secretion of
the bile. We shall soon have occasion to bring
them to the recollection of the reader.

When speaking of organic agents, we noticed
the probability of the opinion, that in living
beings, there exists a series of agencies gradually
raised one above another; each agency having
more or less control over all those below itself.
Now, in the digestive and assimilating functions,
we appear to have, as we might expect, the lowest
of these agencies. The agencies operating in

digestion, and in the first stages of assimilation are, in man, the same perhaps, that exist in all organized beings, vegetable as well as animal ; and are only a few degrees, as it were, above the agencies of mere inorganic matter. This resemblance is inferred from the phenomena of assimilation ; not less than from the peculiar character of the nerves, distributed over the digestive organs ; the effects of which nerves, as we shall presently endeavour to show, approach more nearly to the effects of common chemical agents, than to those of any agent belonging to the animal economy. These nerves compose what, from their peculiar structure, are termed the *ganglionic nerves.* In animals of the very lowest kind, the ganglionic nerves alone appear to exist; and though, in the more perfect animals, the ganglionic nerves are connected with others of a higher character, these nerves always form alone, a peculiar system ; the functions of which seem to be of the subordinate character above noticed.

2. *Of Alimentary Substances.*—It may be considered as a general rule, that organized beings adopt, as aliments, substances lower than themselves in the scale of organization ; or which, if not originally lower, are in some measure lowered, by certain spontaneous changes they undergo. There are, of course, innumerable ex-

ceptions to this rule : but viewing the whole of animated beings, it seems to be a law of nature. Thus plants, and perhaps the very lowest kinds of animals, have the power of assimilating carbonic acid gas : the powers of assimilation of plants, and of such animals, may also extend to other inorganic compounds of carbon—indeed plants and zoophytes appear to derive their chief nourishment from matters of that nature. Higher in the zoological scale, we find that animals almost invariably prey on those inferior to themselves, either in magnitude, in organization, or in intelligence; till we arrive at man himself. He, as his necessities, or as his fancies may dictate, appropriates every nourishing substance, even carbonic acid gas; which the human stomach seems to have the capacity of assimilating, in common, probably, with the stomachs of all animals. Of course a lion, or even a crab, can feed on the body of a man, as well as on that of an ox or of an insect. But no one, we presume, will assert, that man is the natural prey or food of these animals; and such alone is the degree of immunity, for which we here contend : in all the operations of nature, we must try to discover and bear in mind, not the exception, but the rule; otherwise we shall be constantly liable to error.

By this beautiful arrangement in the mode of their nutrition, the more perfect animals are exonerated from the toil of the initial assimilation of the materials composing their frame; since the constituent elements of their food are already in that order which is adapted for their purpose. Hence the assimilating organs do not require the complication, which would otherwise have been necessary; and much elaborate organization is saved. Striking illustrations of this abridgement of organization, are afforded by the differences before mentioned, between the assimilating apparatus of carnivorous and of graminivorous animals. According to the scale which this difference exhibits, we can form some conception of the complication which would have been requisite, to enable such an animal as man to feed, like a plant, on carbonic acid gas; or carburetted hydrogen; or on any other simple compound of carbon.

Another great purpose is effected by this arrangement regarding the food of animals; without which, organization, as at present constituted, could hardly exist. If organized beings did not prey on each other, their remains would accumulate in such quantity, as to be nearly incompatible with life; certainly with animal life in its most perfect condition, as it is at present known to us. But by the arrangement that animals are food to each other; not only is an opportunity afforded, for the existence of a greater number of animals, and of a greater variety among them; but the obtrusion of the bodies of animals, whose life has become extinct, is entirely prevented:

nor, is the removal of dead animal matter the only good accomplished, but many other important results are obtained. To enter upon the consideration of these, would be foreign to our present object. There is, however, one consequence of this system of universal voracity, which more immediately concerns us, since it is of a nature so comprehensive, as to suggest a natural classification of alimentary substances : we allude to the similarity of composition among the staminal principles which constitute the fabric of organized beings.

In our introductory remarks on the chemistry of organization, we showed that organized matters, however apparently dissimilar, yet, chemically speaking, are often nearly related. Of this relation we gave as an example, the composition of the extensive class of substances, denominated the *saccharine group;* all which substances, notwithstanding the endless diversity of their appearance, are essentially alike in their composition, and consist of carbon associated with water. Saccharine substances are chiefly obtained from vegetables ; sugar being the characteristic staminal principle of the vegetable kingdom.

Another well known class of bodies, existing both in vegetables and in animals, are those whose character is *oily.* Oleaginous bodies occur in an infinite variety of forms, some being solid,

others fluid ; yet, in every instance, their peculiar properties are so strongly marked, that we
seldom hesitate about their nature. In this dis-
tinctness of outward appearance, oily bodies are
strongly contrasted with the saccharine group
before mentioned ; many of which group have
few *apparent* and sensible properties in common.
The composition of all the bodies of this oleaginous group, which we have hitherto had an
opportunity of examining in a satisfactory manner, we have found to be essentially the same :
they are either composed of olefiant gas and
water, or have a reference to that composition.
Such is also the composition of the well known
proximate principle termed *spirit of wine*, or *alcohol;* into which principle most substances belonging to the saccharine group, under favourable circumstances, are readily convertible by
the process termed fermentation.

When any part of an animal body, (with the
exception perhaps of entirely oleaginous matters,) is boiled in water, it is separated into two
portions,—one soluble in water, and forming
with the water a tremulous jelly, or *gelatine*—
the other remaining insoluble, indeed becoming
harder, the longer it is boiled ; and from the
identity of its properties with those of white of
egg, denominated *albumen*. Gelatine and albumen exist in very different proportions in the
different textures ; some of these textures, as the

skin, being almost entirely convertible into gelatine; while others yield comparatively little gelatine, and consist principally of albumen. In no animal compound does gelatine exist as a fluid: hence, gelatine has been supposed to be produced by boiling; but the supposition does not appear to be well founded. Gelatine may be considered as the least perfect kind of albuminous matter existing in animal bodies; intermediate, as it were, between the saccharine principle of plants, and thoroughly developed albumen: indeed, gelatine in animals, may be said to be the counterpart of the saccharine principle in vegetables; it being distinguished from all other animal substances, by its ready convertibility into a sort of sugar, by a process similar to that by which starch may be so converted. Albumen exists in the fluid state as a component of the blood: small quantities of fluid albumen are also contained in certain animal secretions: but there is much more of the principle in a solid state; forming what is termed *coagulated* albumen. The blood likewise contains *Fibrin*, another modification of the albuminous principle, in a fluid, or at least in a suspended state: though the most frequent condition of Fibrin, is that of a tough fibrous mass, in which condition, together with albumen, it forms the basis of the muscular or fleshy parts of animals. The *curd* of milk is also a modi-

fication of the albuminous principle. Another
modification of the same principle is the substance called *gluten*; this substance though most
abundant in vegetables, so far resembles the
fleshy parts of animals, as to be, in like manner,
capable of separation into two portions, analogous to gelatine and albumen. Neither of these
modifications of albumen exhibits the quality
possessed by gelatine, of being artificially convertible into saccharine matter ; at least by any
known process : but all of them, including gelatine, differ from the oleaginous and the saccharine principles, in this respect ; that they contain
a fourth elementary principle, namely, azote.
The exact composition of the albuminous group
cannot at present be stated.

Such are the *three great staminal principles*,
from which all organized bodies are essentially
constituted. Of these staminal principles it has
already been remarked, that, without any alteration of their essential composition, they are
capable of assuming an infinite variety of modified forms ; many of which are so peculiar, that
from their sensible properties, it is very difficult
to recognise their identity. Moreover, these staminal principles, in all their forms, are capable
of readily passing into one another, and of combining with each other ; at least the organic
agents, as we shall see hereafter, have the power
of effecting such changes. Further, these sta-

minal principles are all susceptible of trans-
mutation into new principles, according to certain laws: thus the saccharine principle is
readily convertible into the acid, termed oxalic;
or, under other circumstances, into the modification of the oleaginous principle, alcohol.
Though an endless variety of these modifications of the staminal principles exist in different
organized beings, accompanied by numerous
foreign bodies; still the proportion they bear
to the staminal principles is very limited; and
they are either confined to glandular secretions;
or are excrementitious; or extravascular: that
is to say, these modifications and combinations
form no part of the living animal;· though they
are attached to it: as in the case of the various
products of secretion; the shells of the molluscous tribes; and many others.

The consequence then, to which· we before
alluded is; that as all the more perfect organized beings feed on other organized beings, *their*
food must necessarily consist of one or more of the
three staminal principles of organization. Hence,
it not only follows, as before observed, that in
the more perfect animals, all the antecedent
labour of preparing these compounds *de novo*, is
avoided; but that a diet to be complete, must
contain more or less of all the three staminal
principles. Such, at least, must be the diet of
the higher classes of animals, and especially of

man. It cannot indeed be doubted that many animals, on an emergency, have the power of forming a chyle from one of these classes of aliments; but that any of the higher animals can be so nourished for an unlimited time, is exceedingly improbable. Nay, if we judge according to what is known from universal observation, as well as from experiments which have been actually made by physiologists regarding food; we are led to the directly opposite conclusion: namely, that the more perfect animals could not exist on one class of aliments; but that a mixture, of two at least, if not of all the three staminal principles, is necessary to form an alimentary compound well-adapted to their use.

This view of the nature of aliments is singularly illustrated and maintained by the familiar instance of the composition of *Milk*. All other matters appropriated by animals as food, exist for themselves; or for the use of the vegetable or animal of which they form a constituent part. But milk is designed and prepared by nature expressly as food ; and it is the *only material* throughout the range of organization that is so prepared. In milk, therefore, we should expect to find a model of what an alimentary substance ought to be—a kind of prototype, as it were, of nutritious materials in general. Now, every sort of milk that is known, is a mixture of the three staminal principles we have described;

P.   I I

in other words, milk always contains a *saccharine* principle, a *butyraceous* or *oily* principle, and a *caseous*, or strictly speaking, an *albuminous principle*. Though, in the milk of different animals, these three staminal principles exist in endlessly modified forms, and in very different proportions; yet neither of the three is at present known to be entirely wanting in the milk of any animal.

Of all the evidences of design in the whole order of nature, Milk affords one of the most unequivocal. No one can for a moment doubt the object for which this valuable fluid is prepared. No one can doubt that the apparatus by which milk is secreted has been formed specially for its secretion. No one will maintain that the apparatus for the secretion of milk arose from the wishes or the wants of the animal possessing the apparatus; or from any fancied plastic energy. On the contrary, the rudiments of the apparatus for the secretion of milk must have actually existed in the body of the animal, ready for developement, before the animal could have felt either wants or desires. In short, it is manifest that the apparatus and its uses, were designed, and made what they are, by the great Creator of the universe ; and on no other supposition can their existence be explained.

The composition of the substances, by which animals are usually nourished, favours the mixture of the primary staminal alimentary princi-

ples; since most of these substances are compounds, of at least two, of the staminal principles.
Thus, most of the gramineous and herbaceous
matters contain the saccharine and the glutinous
principles; while every part of an animal contains at least albumen and oil. Perhaps, therefore, it is impossible to name a substance constituting the food of the more perfect animals,
which is not essentially a natural compound of
at least two, if not of all the three great principles of aliment. But in the artificial food
of man, we see this great process of mixture
most strongly exemplified. He, dissatisfied with
the spontaneous productions of nature, culls from
every source ; and by the force of his reason,
or rather of his instinct, forms in every possible
manner, and under every disguise, the same
great alimentary compound. This after all his
cooking and his art, how much soever he may
be disinclined to believe it, is the sole object
of his labour ; and the more nearly his results
approach to this object, the more nearly do they
approach perfection. Even in the utmost re-
finements of his luxury, and in his choicest
delicacies, the same great principle is attended
to ; and his sugar and flour, his eggs and butter,
in all their various forms and combinations, are
nothing more or less, than disguised imitations
of the great alimentary prototype MILK, as furnished to him by nature.

484

CHAPTER III.

OF THE DIGESTIVE PROCESS ; AND OF THE GENERAL
ACTION OF THE STOMACH AND DUODENUM.

WE proceed now to consider the most important function of the stomach, by which the assimilation of the food is begun.   But before that function can be well understood, it is necessary to make a few remarks on the influence of water, in modifying the intimate constitution and the peculiar properties of alimentary substances. We have intentionally delayed these remarks, in order that in this place the chemical influence of water might be more strikingly exemplified.

Water enters into the composition of most organized bodies in two separate forms; which must be clearly distinguished, and which it is requisite that the reader should always bear in mind.   Water may constitute an *essential* element of a substance, as of sugar or of starch in their dryest states ; in which case, the water cannot be disunited without destroying the compound : or water may constitute an *accidental* ingredient of a substance, as of sugar or of starch in their moist states ; in which case more or less of the water may frequently be removed, without destroying the essential properties of the compound. Now, a very large number of organized bodies, (perhaps all those to which our present enquiry relates) contain water in both these forms ; both as an essential element, and as an accidental ingredient ; and in most instances, it is impossible to discriminate between the water which is essential, and that which is accidental. The mode of union, however, among the elements of bodies, in these two states of their combination with water, must be altogether different. Wherein the difference consists, is very imperfectly known ; but, from the explanation we shall now offer, the reader will more fully understand the nature of these two modes of union : perhaps some light may even be thrown on the cause of their difference.

In the first part of this volume, we stated that the molecular, or combining, weights of carbon and of water are, by chemists, usually considered to be represented by the numbers 6 and 9 ; the weight of hydrogen being *one*. We also advanced the opinion, that the molecules or atoms of carbon and of water, where more than one exist, instead of remaining separate, as is now supposed, are associated together into groups, or supermolecules ; and that carbon, water, and similar bodies, always enter into combination, not as single molecules, but as one supermolecule. To illustrate our meaning, let us take as examples, the state of combination of the molecules constituting the different varieties of sugar.

Sugar from the cane, in its purest state, and when as free as possible from accidental water, is, according to the present language of chemists, composed of 9 atoms of carbon and 8 atoms of water. Now, we suppose these 9 atoms of carbon, and 8 atoms of water, to be associated into two supermolecules, weighing (9×6) 54, and (8×9) 72, respectively. So that, we conceive a molecule of sugar from the cane to be a binary compound, of a supermolecule of carbon weighing 54, and a supermolecule of water weighing 72. Again, the sugar of honey, according to the present language of chemists, is composed of 9 atoms of carbon, and 12 atoms of water; or, according to our view of molecular arrangement, the sugar of honey is composed of two supermolecules, one of them, carbon, weighing 54, as in the sugar of the cane—the other, water, weighing no less than (12×9) 108. A similar statement may be given of the composition of Lignin, another of the saccharine class of bodies. Lignin, which, in all its various forms appears to consist essentially of equal weights of carbon and water, may be said to be composed of 9 atoms of carbon, and 6 atoms of water; or, according to our views, of two supermolecules weighing (9×6) 54, and (6×9) 54, respectively.

Hence, the saccharine class of bodies may be represented in the following manner:—

Carbon.		Water.	
54	+	54	Lignin.
54	+	72	Cane Sugar; Wheat Starch.
54	+	108	Sugar of honey; Arrowroot.

The molecular constitution of the saccharine bodies, above stated, may be compared with the molecular constitution of Vinegar. According to the present language of chemists, vinegar, in its purest and most detached form, is composed of 4 atoms of carbon, and 3 atoms of water; or, according to our views, of two supermolecules weighing (4×6) 24, and (3×9) 27, respectively; while crystallized vinegar contains the same proportion of carbon with one-third more of water. Thus, the molecular constitution of these two different forms of vinegar may be represented as follows:—

Carbon.		Water.	
24	+	27	Absolute vinegar.
24	+	36	Crystallized or solid vinegar.

We have stated the composition of vinegar, in order to draw the attention of the reader, to the difference between the supermolecule of the carbon in that acid, and the supermolecule of the carbon in the saccharine class of bodies; a difference to which these two classes of bodies probably owe the striking differences in their sensible properties. But why the supermolecule of carbon should be 54, in bodies of the saccharine class, and why this supermolecule should in general exist in the self-attractive form, and produce sweetness; or why the supermolecule of carbon in vinegar should be 24, and why this supermolecule should have such a tendency, as it exhibits, to assume the self-repulsive form, and to produce sourness; we do not know, and probably shall never be able fully to explain. Still, there can be little doubt that a careful and philosophical examination of the phenomena, would go far to dispel the obscurity in which the subject is now involved.

Such are the principles, which, we conceive to regulate the chemical union of organic, and indeed of all other compounds; and if chemical union be so regulated, the inferences are most curious and important. With these inferences in general, we have at present no concern: but the inferences more particularly deducible from alimentary compounds are the following:—

First. We would draw the attention of the reader to the contrast between the two supermolecules of carbon, and of water, constituting sugar; the supermolecule of carbon being uniform throughout the whole saccharine class, while the supermolecule of water is that which is variable. Now, there is reason to believe that this contrast holds in other instances; that in different organized substances of the same kind, the supermolecule of carbon, or of some of its compounds, remains the permanent and *characteristic* element; and that the different modifications are produced by variations in the supermolecule of water; which may be called the *modifying* supermolecule.

Secondly. The manner of the operation of the modifying agency may be thus illustrated. If to a portion of cane sugar, we add that quantity of water, which, by an easy calculation, we learn is necessary to be united with cane sugar, in order to its conversion into sugar of honey; we find that we cannot succeed in producing such conversion; and that the excess of water which had been added, flies off, and leaves the cane sugar in its original state. On the other hand, if we apply heat to the sugar of honey; though we may indeed drive off part of the water essentially associated with that sugar, we do not obtain sugar similar to cane sugar; but we destroy, or altogether decompose the sugar of honey. These facts, therefore, show that the excess of water, constituting the difference of the sugar of honey from the sugar of the cane, is really in some state of essential union, incapable of being imitated; while, in the cane sugar, the water may exist as an accidental ingredient only. In truth, according to our views of molecular arrangement; *every individual supermolecule of the weaker sugar contains a portion of this excess*

of water, as an essential element of its composition.
Hence such water cannot be separated from any
compound, without destroying the entire crasis,
or constitution, of its molecular elements; which,
as in the case of the sugar of honey, we find, by
experiment, to be the result. On the other hand,
we suppose that the molecules of accidental
water, *form no essential element of the molecules
of sugar, or of other bodies; but that these accidental molecules of water, are only in a state of
loose association with the essential molecules of
sugar or of other bodies;* and hence, the ease
with which accidental water may be separated
without destroying such bodies.

Thirdly. It may be advanced as a general
rule, that the larger the number, representing
the weight of the supermolecule of any compound substance; whether such number represent the characteristic, or the modifying supermolecule; the more easily may that compound
substance be decomposed. Thus, the sugar of
honey is more easily decomposed—is much less
permanent, than the sugar of the cane; and the
purest sugar, is much less permanent than Lignin. In like manner, when water is the modifying element of any compound, as it is in most
organic compounds; the larger the number representing the supermolecule of the water; the
greater, for the most part, is the solubility of the
compound.

Fourthly. There are at present no chemical terms corresponding to those differences of composition, which we have brought under the notice of the reader. Now, the terms *strong* and *weak,* which in commerce, distinguish the different varieties of sugar, are sufficiently expressive ; we have, therefore, made choice of them, to denote the similar varieties of other organic compounds. Thus, when we speak of a *strong* compound ; we mean that its constituent supermolecules are, like those of strong cane sugar, less complicated than the supermolecules of a *weak* principle, like those of the sugar of honey. Again, there are no terms expressive of the conversion of a strong substance into a weak substance, or the contrary. To express such conversion we have adopted the terms *reduction,* and *completion.*

In the above illustrations of the modifying influence of water in organic compounds, we have selected sugar as our example ; solely from its being the most familiar. But, as we have more than once noticed, exactly the same laws appear to regulate the composition of all organized bodies. Thus in the *strong,* fixed, and solid, oils or fats, the characteristic supermolecule of which, as we have already said, has some relation to olefiant gas ; the modifying molecule of water is very small, perhaps, in some oleaginous bodies, is even a submolecule. Whereas, in alcohol, which is the *weakest* condition of the oily principle ; the weight of the modifying supermolecule of water is more than half that of the olefiant gas, and alcohol is perfectly soluble in water.

Gelatinous and albuminous substances, also, exhibit precisely the same variations. The *strong* tenacious glue, employed in the arts, is made from the firmer parts of the hides of old animals; while the gelatinous size, or weak glue, is made from the skins of younger and more delicate animals. These two varieties of glue differ from one another, in the weights of the modifying supermolecules of water which enter into their composition. In general, it may be observed, that the substances composing the frame of old and of young animals, differ chiefly in the weights of their modifying supermolecules of water ; and that the dissimilarity of their properties, is chiefly owing to this difference.

If the reader has clearly apprehended, and will bear in mind, the principles which have now been stated, as regulating the chemical constitution of organized bodies, and the modes in which organized bodies are influenced by their modifying constituent, water ; he will be able to accompany us in the observations we are about to offer ; and he will thus, more especially, be able to form a general conception of the *chemical* operations of the stomach.

The operations of the stomach, viewed as a whole, may be stated as follows:—

1. The stomach has the power of dissolving alimentary substances, or, at least of bringing them to a semifluid state. This operation seems to be altogether chemical; and is probably effected by *reducing* the properties of alimentary substances.

2. The stomach has, within certain limits, the power of changing into one another, the simple alimentary principles, which have been described in the last chapter. Unless the stomach possessed such a power; that uniformity in the composition of the chyle, which we may imagine to be indispensable to the existence of every animal, could not be preserved. This part of the operations of the stomach, appears, like the reducing process, to be chemical; but not so easy of accomplishment: it may be termed the *converting* operation of the stomach.

3. The stomach must have, within certain limits, the power of *organizing* and *vitalizing* the different alimentary substances; so as to render them fit for being brought into more intimate union with a living body, than the crude aliments can be supposed to be. It is impossible to imagine, that this organizing agency of the stomach can be chemical. This agency is *vital*, and its nature is completely unknown.

1. *Of the Reducing Powers of the Stomach.*—
In order to render more intelligible that function
of the stomach, which is performed by means of
its reducing power; let us endeavour to trace
the series of phenomena, which appear to arise
during the conversion of simple albuminous matter into the albumen of the chyme; without
taking into account any other change.

When a portion of fluid albumen, like the
white of an egg, or milk, is introduced into
the stomach of an animal, as of a dog; it instantly becomes solid; or, in ordinary language,
is *coagulated.* This coagulation is probably a
mere chemical change; for the same change,
would, under similar circumstances, take place
out of the body. That is to say; if the white of
an egg, or milk, were mixed with a fluid more or
less acid, like the fluid which exists in the stomachs of animals while the food is undergoing
the process of digestion; the white of egg, and
milk, would be coagulated. There may be,
however, and probably is, some object in the
change produced by coagulation; since the stomachs of animals are fitted to operate chiefly on
solid matters. Admitting the object of the
change, we can hardly consider it to be essential
to the subsequent processes; for gelatine, a
staminal alimentary principle, nearly resembling
albumen in its composition, undergoes, under
similar circumstances, no such solidifying

change. The albumen which has by the coagulating influences of the juices of the stomach,
been solidified into a mass or curd, is soon
altered further ; more especially that part of the
mass, immediately in contact with the membrane
of the stomach.   The curdy mass assumes a
gelatinous appearance ; then each portion is
successively more and more softened ; till at
length, the whole becomes nearly fluid, and
after some additional modifications, gradually
passes into the state of chyme.   Through all
these apparent changes, however, the albumen
has undergone no real change.   What was
introduced into the stomach as albumen, is still
albumen in the chyme ; at least chemists have
pronounced it to be so.   Yet the albumen has
assumed an appearance altogether different.
The albumen of the egg, out of the stomach,
may be coagulated by heat, into a firm and
elastic solid.   The albumen of the chyme is
indeed coagulable by heat ; but its coagulation
is so imperfect, and so wanting in tenacity, as to
offer a striking contrast with the coagulated albumen of the egg.   What then, in the stomach,
has happened to the albumen ?   Viewing only
its susceptibility of coagulation, the albumen
has merely become chemically combined with
a portion of water.   The solid and tenacious
albumen has, by this combination with water,
been reduced to the *weakest* possible state—to

the delicate state, as it were, of infancy; in short, to a state precisely analogous to that of the weak sugars, and other organic compounds, comparatively with the strong and perfect varieties of the same substances; as described in the preceding chapter.

Such is, we believe, an accurate account of the merely solvent or reducing powers of the stomach. We have next to show the means by which this solution or reduction is effected.

The process of combining different substances with water, and of thus reducing them from a stronger to a weaker condition, may, in some instances, and to a certain degree, be effected artificially. But in no instance do we appear to be able to invert the process; or to complete an organic compound, by separating the compounds from the water which enters into its composition. For example, we can, in some respects, make a strong sugar weak, but we cannot change a weak into a strong sugar; though such a change, within certain limits, seems to be, to the organic agents, just as easy, as the reducing process.

The different operations of cookery, as roasting, boiling, baking, &c. have all a reducing effect; and may, therefore, be considered as preparatory to the solvent action of the stomach. Of these operations, Man's nature has taught him to avail himself, and they constitute the

chief means by which he is enabled to be omnivorous: for, without such preparation, a very
large portion of the matters which he now
adopts as food, would be completely indigestible.
By different culinary processes, the most refractory substances, can often be rendered nutritious.
Thus, by alternate baking and boiling, the
woody fibre itself may be converted into a sort of
amylaceous pulp; not only possessing most of
the properties of the amylaceous principle, but
capable of being formed into bread. The culinary art engages no small share of attention
among mankind; but, unfortunately, cooks are
seldom chemists; nor indeed do they understand the most simple of the chemical principles
of their art. Hence, their labour is most frequently employed, not in rendering wholesome
articles of food more digestible, which is the true
object of cookery; but in making unwholesome
things palatable; foolishly imagining that what
is agreeable to the palate, must be also healthful
to the stomach. A greater fallacy can scarcely
be conceived; for, though by a beautiful
arrangement of Providence, what is wholesome
is seldom disagreeable; the converse is by no
means applicable to man; since those things
which are pleasant to the taste, are not unfrequently very injurious. Animals, indeed, for
the most part, avoid, instinctively, all unwholesome food; probably because every thing that

P.   K K

would be prejudicial, is actually distasteful to them. But as regards man, the choice of articles of nourishment has been left entirely to his reason.

In order to illustrate the importance of a judicious adaptation of cookery; we may observe, that the particular function of the stomach, now under consideration, namely, the dissolving or reducing function, is liable to very great derangements. In some individuals, the reducing power is so weak, that their stomach is almost incapable of dissolving solid food of the most simple kind. In such a state of the stomach, a crude diet of the flesh of animals in a hardened state, or of other compact substances, is little else than poisonous; while the same animal and vegetable matters often agree well, if reduced to a pulpy state. On the other hand, as in the disease termed Diabetes, the solvent powers of the stomach are often inordinately increased; and every article of food is dissolved and absorbed almost as soon as it is swallowed. In such cases, a diet and a mode of preparation are required, directly the reverse of those which are found to be so beneficial, when there is a debility of the solvent powers; and aliments must be chosen which are firm and solid, but at the same time nutritious.

Regarding the intimate nature of the agency, by which the combination of alimentary substances with water is effected in the stomach, we cannot be said to possess much certain knowledge. This combination appears to be chiefly owing to the agency of a fluid, secreted by the stomach; the glands for the formation of which fluid, are most numerous toward the pyloric orifice. The aliment having been previously broken down by mastication, and having received an admixture of saliva and of other fluids, is brought into contact with the fluid secreted by the stomach: by this secretion of the stomach, or by some other energy exerted in that organ, the food which has been introduced into the stomach is associated with water; and thus becomes itself more or less a fluid. Of this important secretion of the stomach, *chlorine*, in some state or other of combination, is an ingredient: it would seem a necessary ingredient; for the secretion in its healthy state, always contains more or less of chlorine; the powerful influence of which elementary principle, seems mainly to contribute towards effecting the union of the food with water. The chlorine, thus so indispensable to the reducing process, is perhaps more frequently the subject of derangement, than any thing concerned with the assimilation of the food. It often happens that instead of chlorine, or a little free muriatic acid, a large quantity of free muriatic acid is elicited; which free muriatic acid not only gives rise to much secondary uneasiness, but more or less retards the process of reduction itself. The source of this chlorine or muriatic acid, must be the common salt which exists in the blood : to suppose that it is generated, is quite unnecessary. The chlorine is therefore secreted from the blood ; and it may be demanded, what is the nature of the agency, capable of separating chlorine from a fluid so heterogeneous as the blood ? We are acquainted with one agent that exerts such a power ; namely electricity : and this agent, as we formerly observed, seems to be employed by the animal economy for its operations, in the same manner, and on the same principles, as the materials themselves are employed, from which the animal body is constructed. Perhaps, therefore, the decomposition of the salt of the blood may be fairly referred to the immediate agency of this principle, electricity. But here the question arises—What becomes of the soda from which the muriatic acid has been disunited? The soda remains behind, of course, in the blood, and a portion of it, no doubt, is requisite to preserve the weak alkaline condition essential to the fluidity of the blood. But the larger part of this soda is probably directed to the liver, and is elicited with the bile in the duodenum ; where it is thus again brought into union with the acid, which had been separated from the blood by the stomach. These observations, illustrating the importance of common salt in the animal economy, seem to explain, in a satisfactory manner, that instinctive craving after this substance, which is shown by all animals.

Admitting that the decomposition of the salt of the blood is owing to the immediate agency of galvanism : we have, in the principal digestive organs, a kind of galvanic apparatus ; of which the mucous membrane of the stomach, and perhaps the mucous membrane of the intestinal canal generally, may be considered as the acid or positive pole ; while the hepatic system may, on the same view, be considered as the alkaline or negative pole. Whether such galvanic action be admitted or not ; and the admission is of no very great importance ; what we have above stated may be received as a simple expression of the facts, in so far as they relate to the saline constituents of the blood. Moreover, be the nature of the energies what they may, by which these changes are effected ; along with these changes, and probably by the aid of the same energies, other very important changes or processes are carried on, to some of which we shall presently have occasion to allude. In the mean time, we may close this account of the preliminary function of the stomach, by noticing the strong grounds there are to believe, that the solvent power, which we have described, or some power having a great resemblance to it, exists

not only in the stomach, but in every part of an
animal body. In all animals there are minute
tubes, called absorbents, which originate in every
part of their bodies, and at length uniting, enter
the sanguiferous system along with the chyle.
Now, the office of these tubes, is to remove all
the portions of the animal frame, which after
having performed their several functions, require
to be withdrawn. Of course, before solid parts
can be thus removed, they must be dissolved,
(*digested* in fact); and such solution, in many
instances, is probably effected, as it is in digestion, by combining these solid parts with water.
This supposed analogy between the solvent
actions of the stomach, and those actions which
must prevail all over the body, seems to be
strongly confirmed by that similarity of structure
and of function existing between the lacteals
and the absorbents: they indeed form but one
system. We shall resume this subject hereafter.

2. *Of the Powers of Conversion possessed by the
Stomach.*—Though the proportions of the different ingredients of the chyle, as ultimately
formed, are liable to be much varied, according
to the nature of the food ; yet, whatever the nature of the food may be, the general composition
and character of the chyle, remain always the
same. The stomach must, therefore, be endowed

with a power or faculty, the agency of which is
to secure this uniform composition of the chyle,
by appropriate action upon such materials as
circumstances may bring within its reach. Two,
indeed, of the chief materials from which chyle
is formed, namely, the albuminous and the oleaginous principles, may be considered to be
already fitted for the purposes of the animal
economy, without undergoing any essential
change in their composition. But the saccharine class of aliments, which form a very large
part of the food of all animals, except of those
subsisting entirely on flesh, are by no means
adapted for such speedy assimilation. Indeed,
one or more essential changes must take place
in saccharine aliments, previously to their conversion, either into the albuminous, or into the
oleaginous principles. Most probably, under
ordinary circumstances, these essential changes
are altogether chemical; that is to say, these
changes are such as do take place, or rather,
such as would take place, if the elements of the
substances thus changed in the stomach, could,
out of the body, be so collocated, as to bring
into action the affinities necessary for the
changes produced in the stomach. Thus, as
we know, the saccharine principle spontaneously
becomes alcohol; which, as has been stated, is
merely an oleaginous body of a weak kind.
When, therefore, in the stomach, it is requisite

that sugar be converted into oil; it is probable that the sugar passes through precisely the same series of changes, it undergoes, out of the body, during its conversion into alcohol. We cannot trace the conversion of sugar into albumen; because we are ignorant of the relative composition, and of the laws which regulate the changes, of these two substances. The origin of the azote in the albumen, is likewise at present unknown to us; though in all ordinary cases, it seems to be appropriated from some external source. That the oleaginous principle may be converted into most, if not into all the matters necessary for the existence of animal bodies, seems to be proved by the well-known fact, that the life of an animal may be prolonged, by the absorption of the oleaginous matter contained within its own body. Thus, many hybernating animals, when they retire in autumn, to sleep during the winter, are enormously fat. But while they sleep, their fat is gradually removed; till they awake in the spring quite divested of it, and in a state of inanition.

The reader will have remarked that we have made use of the term *ordinary circumstances;* and perhaps it may be not amiss to explain what meaning we attach to that term.

When an animal is duly fed according to that diet which is natural to it, and for which its organization has been adapted: a regular and

ordinary series of changes takes place within
the animal, and the alimentary matters are converted into chyle. But one general characteristic of organized beings is, that within certain
limits, and for a certain time, they possess the
power of varying their habits, and of accommodating themselves to circumstances. Under
extraordinary circumstances, therefore, extraordinary changes must, and do, take place. In
some instances, these changes out of the ordinary
course, are to an extent altogether astonishing ;
and such as defy our utmost calculation. The
assimilating organs appear even to decompose
principles which are still considered as elementary ; nay, to form azote or carbon ; so that
it is impossible to define what, on an emergency,
these organs are capable of doing. But what is
thus done by these organs on an emergency,
will, usually, be found to constitute an exception
to what they do in ordinary ; their ordinary
mode of action being always that which is most
simple, and which is thus to be considered as
the rule.

3. *Of the Organizing and Vitalizing Powers
of the Stomach.*—In this part of our investigation, we meet the real difficulties we have to
overcome in explaining the operations of living
beings. The whole of the great and essential
changes which alimentary substances undergo,

may, and perhaps will, be traced, by care and attention; but all beyond, will probably remain for ever unknown to us. Now at least, though we understand, in some degree, the chemical changes; of the vitalizing influence, we in truth know absolutely nothing. There is, however, every reason to believe that vitality is imparted through the agency of the living animal itself. For though, from the natural composition of alimentary substances, they be, to a certain extent, fitted for the purposes of the animal economy; yet, alone, they are incapable of uniting themselves with the animal frame; and unless the living economy contribute likewise its share of the labour, the future work of assimilation will be incomplete.

Of the Changes the Food undergoes in the Duodenum.—We alluded in general terms to the bile and the pancreatic fluids, when we were treating of the organs by which these fluids are secreted. We have now to consider, more particularly, the nature of these secretions, and their share in the performance of the functions of the duodenum.

With the yellow colour, and the intensely bitter taste of the *bile*, all are familiar: we need not, therefore, dwell on the sensible properties of the secretion, but proceed to notice its chemical composition. The chemical composition of the bile is very heterogeneous, though not perhaps so heterogeneous as has been represented ; since it is probable that many of the ingredients said to be contained in the secretion, are products which have resulted from the methods employed in its analysis. Bile, like all animal fluids, is composed essentially of water ; but the solid matters contained in the bile, are nearly altogether formed from one or more proximate principles, in which carbon and hydrogen predominate. These proximate principles exist simultaneously, if not in conjunction, with soda, with various salts of soda, and with other substances. The properties of the bile vary somewhat in different animals ; but in all animals its essential characters remain wonderfully similar.

We are much less acquainted with the properties of the *pancreatic fluid*, than with the properties of the bile. The pancreatic fluid was formerly supposed to be of nearly the same composition as the saliva ; but recent observations have shown that the pancreatic fluid contains albumen, and a curdy substance. The pancreatic fluid is, for the most part, in a slight degree acid, and holds in solution matters of a saline nature, closely resembling those found in all animal fluids.

When the food which has undergone the first process of digestion in the stomach, quits that organ, and enters the duodenum ; some other changes of a very remarkable kind take place.

If the food originally contained no albuminous matter; no albumen is developed in the stomach; but immediately on the entrance of the semi-fluid mass into the duodenum, and its mixture with the bile and the pancreatic fluids; albuminous, and other chylous matters, become distinctly perceptible. At the same instant, those fluid parts, which in the stomach were acid, are so far altered, by the addition of the bile, and the pancreatic fluids, as to become neutral, or almost neutral: some gas is frequently extricated; and that portion of the food which is destined to be excrementitious, is evidently separated. The albumen, which is thus found to exist in the *chyme*, may be partly derived from the pancreatic fluid, which, as we have already mentioned, has been said to contain albumen. But the quantity of albumen, and of other proximate principles of the chyle, that are found in the contents of the duodenum, at some distance onward from the pylorus, is much too great to be explained in this manner. Indeed, the properties, as well as the quantity, of the albuminous matters, show, beyond a doubt, that the albuminous matters are developed from the food, and constitute the chyle which is subsequently taken up by the lacteals.

Such are those most interesting, and at the same time, most obvious, phenomena, observed in different animals, in which the changes produced on the food by the action of the duodenum have been examined. These phenomena appear to vary considerably, according to the nature of the food : but so far as we can understand the phenomena ; under every change of food, the essential character of the changes which the food undergoes in the duodenum, remains unaltered. That is to say : the acid formed in the stomach, combines, in the duodenum, with the alkali of the bile ; the albuminous principles are developed ; and the excrementitious matters are, more or less perfectly, separated. Of the nature of the more recondite and vitalizing changes which take place in the duodenum ; we are in the same state of complete ignorance, as we are of the similar changes which take place in the stomach ; and probably shall long so remain.

In the preceding remarks on the different processes which take place in the stomach and duodenum, and which are necessary for the conversion of the food of an animal into the living material of its body ; we have endeavoured to distinguish between what, to a certain extent, is within our powers of comprehension, and what is completely beyond them. It remains to be observed in conclusion, that though the three great and essential processes of digestion, namely, the reducing, the converting, and the organizing processes are sufficiently distinct from each other ; yet it is not to be understood that they take place in succession, or in the order in which they have been described. The fact is, all these processes go on at the same time ; and as soon as a portion of food begins to be dissolved, its future changes seem to be determined. If it be necessary that the portion of food undergo an essential change ; that change is accordingly begun. If no such change be required ; the organizing process itself begins simultaneously with the reducing process. The consequence of this union of the digestive processes is, as we have stated, that the staminal principles are all developed in the chyle ; as soon as the excrementitious matters are separated by the biliary and pancreatic fluids.

4. *Of the Functions of the Alimentary Canal, beyond the Duodenum.*—Compared with the functions of the stomach and duodenum, the functions of the succeeding portions of the alimentary canal, as far as we can judge, are unimportant. The digested mass passes from the duodenum into the jejunum, and ilium ; though before the food reaches the end of the ilium, the whole of the chyle contained in it, has been absorbed into the apertures of the numerous tubes named *lacteals.* These tubes open into the whole interior surface of the three portions of the alimentary canal, along which the food is moved from the stomach to the colon : but the lacteals

are most numerous in those parts of the canal nearest to the stomach. From the ilium, the undigested or excrementitious matters proceed into the cœcum; in which cavity, in some animals, as for example, in the horse, even these excrementitious matters appear to undergo a second digestion; but in all animals, the contents of the cœcum have a very different aspect from the contents of any part of the alimentary canal nearer to the stomach. The mass of excrementitious matters continue their course from the cœcum into the colon, where they are still further changed. The nature of these changes, however, is not well understood, though they are probably of no small importance in the animal economy. Finally, all the nutritious portions of the food, having entered into the system of the animal; nothing remains but what is entirely excrementitious.

Such is a short sketch of the phenomena of digestion and assimilation, in so far as these processes are effected by the stomach and the alimentary canal. The phenomena suggest the following reflections:

First. With regard to the nature and the choice of aliments, and the modes of their culinary preparation; it follows from the observations we have offered; that, under similar circumstances, those articles of food which are the least organized, must be the most difficult to be assimilated : consequently, that the assimilation of crystallized, or very pure substances, must be more difficult than the assimilation of any others. Thus, pure sugar, pure alcohol, and pure oil, are much less easy to be assimilated, than saccharine matter in the modified amylaceous form ; or than that peculiar condition or mixture of alcohol existing in natural wines; or than butter. In these modified forms, the assimilation of the saccharine and the oleaginous principles is comparatively easy.   Of all crystallized substances, pure sugar is perhaps the most easily assimilated; but every one is taught by experience, that much less can be eaten of articles composed of sugar, than of articles composed of amylaceous matter.   In some varieties of the disease termed dyspepsia, the effect of pure sugar is most pernicious ; perhaps fully as pernicious as the effect of pure alcohol. Nature has not furnished either pure sugar or pure starch ; and these substances are always the results of artificial processes, more or less elaborate ; in which, as in many of the processes of cookery, man has been over-officious ; and has studied the gratification of his palate, rather than followed the dictates of his reason.   In many dyspeptic individuals, the assimilating and preservative powers of the system, are already so much weakened, as to be unable to resist the crystallization of a portion of their

circulating fluids. Thus in gouty invalids, how
often do we see chalk-stones formed in every
joint? Now, with so little control over their
own fluids, how can they reasonably hope to
assimilate extraneous crystallizations? If, therefore, such an invalid, on sitting down to a luxurious modern banquet, composed of sugar, and
oil, and albumen, in every state and combination, except those best adapted for food, would
pause a moment, and ask himself the question ;
Is this debilitated and troublesome stomach of
mine, endowed with the alchemy requisite for
the conversion of all these things into wholesome
flesh and blood? He would probably adopt a
simpler repast, and would thus save himself
from much uneasiness. The truth is, many of
the elaborate dishes of our ingenious continental
neighbours, are scarcely nutritious, or designed
to be so. They are mere vehicles for different
stimuli—different ways, in short, of gratifying
that low animal propensity, by which so many
are urged to the use of ardent spirits, or of various narcotics. In one respect, indeed, namely,
in reducing to a state of pulp, those refractory
substances which we have before mentioned, the
culinary processes of our neighbours are much
superior to ours ; but in nearly every other
respect, and most of all, in the general use of
pure sugar and pure oil, their cookery is emi-
nently injurious to all persons who have weak

digestion. On the other hand, in this country, we do not in general pay sufficient attention to the reducing processes of the culinary art. Every thing is firm and crude; and though the mode of preparation be less captivating; the quantity of indigestible aliment is quite as great in our culinary productions, as in those of France.

We are not, however, writing a treatise on cookery or dietetics: but in treating of the function of digestion, it is impossible altogether to pass over these important subjects. The foregoing observations are merely intended as illustrations of those general principles which often regulate the choice, and the preparation of the food of mankind, in a state of civilized society. Reason is too little followed, the indulgence of the palate is the sole object; so that the organs of digestion already enfeebled, and incapacitated for the assimilation, even of the most proper nourishment, are daily oppressed with a task for which they are altogether unequal. The consequence is, that though for a time the labour be sustained, the digestive energies are at length overcome. The dyspeptic being passes half his days in misery; his offspring inherit their parents' constitution; and if they persist in a like course of slow poison; after a few generations, the race becomes extinct,—" his name even is cut off from among men!" Providence has gifted man with reason; to his reason, therefore, is left the choice of his food and drink, and not to instinct, as among the lower animals: it thus becomes his duty to apply his reason to the regulation of his diet; to shun excess in quantity, and what is noxious in quality; to adhere, in short, to the simple and the natural ; among which the bounty of his Maker has afforded him an ample selection; and beyond which if he deviates, sooner or later, he will suffer the penalty.

Secondly. The view we have now taken of the processes of digestion, removes in some degree that mysterious character with which they have been invested ; and by lessening the field of our enquiry, brings us nearer to our object. We had previously known, that the articles employed as food by animals, are essentially composed of three or four elements. But we have now learnt, that all the more perfect of those matters on which animals subsist, are compounds of three or four proximate principles ; the whole of which compounds, except one, (the saccharine), are, in their essential characters, identical with the materials composing the frame of the animals themselves. We have also learnt, that owing to this identity of composition, many animals are saved the labour of forming these proximate principles from their elements; and have only to re-arrange them, as their exigencies may require. The task of forming the proximate principles is thus left to the inferior animals or to plants; which are endowed with the capacity of compounding these proximate principles from matters still lower in the scale of organization, than the animals and plants themselves. Hence there is a series, from the lowest being that derives its nourishment from carbon and carbonic acid, up to the most perfect animal existing. Each individual in the series preferring to assimilate other individuals immediately below itself; but having, on extraordinary occasions, and in a minor degree, the power of assimilating all, not only below, but above itself, in the system of organized creation.

Thirdly. We stated that the immediate influence employed by the organic agent is probably galvanism, or the common agents operating among inorganic matters ; and that the digestive apparatus, viewed as a whole, seems to be arranged on galvanic principles. We wish, however, our readers clearly to understand, that we consider the organic agent residing in the ganglionic system of nerves, and employing the electric agency, to be *not* electricity itself; though the ganglionic agency is probably the lowest kind existing in animal bodies, and only, as it were, one degree above the agencies of inanimate matter. We dwell on this point the more, because from deficient recollection of what electricity is, and what are the living powers acting through the nervous system of animals ; it has been maintained, nay, has even been endeavoured to be experimentally proved, that these nervous powers are identical with the powers of electricity. It is impossible to imagine a greater fallacy. Admitting that electricity, properly directed, could change the proximate elements of the food into those of chyle; can we imagine electricity to vary spontaneously its mode of operation, so as to produce the same chyle from every sort of aliment—that electricity is an *intelligent* agency acting with a certain object? Besides, if the nervous agency be identical with electricity in one set of nerves, it must be more or less identical with electricity in all; for though powers of a higher order may be imagined to reside in different classes of nerves, the whole nervous system must be supposed to possess, in common, certain other powers, analogous to, if not identical with, the inferior power residing in the ganglionic nerves: otherwise that free communication, so plainly indicated by the structure, could not be supposed to take place among the different parts of the nervous system. Now, on the supposition that the inferior power residing in the nervous system, be identical with electricity, how different must be the functions of that agency in the different classes of nerves; in one class of nerves, for example, digesting and assimilating the food; in another class helping to convey sight or sound; in the brain itself, shall we say, actually thinking, or at least conveying thought! As to the experiments, on which it has been attempted to rear this most untenable opinion, they prove nothing whatever; and are easily explained on other principles. Such explanation would be foreign to our present object, were we to introduce it here. But there is one observation, which has always appeared to us conclusive against this fancied identity of the nervous energy with electricity; and with which, we shall close what we have to offer, regarding the present subject. Most persons are aware that there are certain fishes endowed with the power of evolving electricity, and of communicating a smart shock to other animals. Now, in all the fishes in which this power resides, as in the Torpedo, there is a complicated apparatus, extending over a large portion of the fish's body, expressly for the purpose of forming the electricity, which the fish communicates; thus, proving beyond a doubt, that mere nerves are not sufficient to develope electricity; and that, when electricity is wanted, an express and peculiar organ is as requisite for its secretion or formation, as for the secretion and formation of any other product of the animal economy.

The further reflections suggested by the facts we have now detailed, will be given in conjunction with the reflections suggested by the facts to be detailed in the next chapter.

CHAPTER IV.

OF THE PROCESSES OF ASSIMILATION SUBSEQUENT
TO THOSE PROCESSES WHICH THE FOOD UNDER-
GOES IN THE STOMACH AND ALIMENTARY CANAL ;
PARTICULARLY OF THE CONVERSION OF THE
CHYLE INTO BLOOD. OF RESPIRATION, AND ITS
USES. OF SECRETION. OF THE FINAL DECOM-
POSITION OF ORGANIZED BODIES. GENERAL RE-
FLECTIONS, AND CONCLUSION.

1. *Of the Passage of the Chyle from the Alimentary Canal into the Sanguiferous System ; and
of the Function of Absorption generally.*—The
Chyle, as we have already said, is taken up from
the alimentary canal, by numerous minute
tubes, named *lacteals;* these tubes being part of
the system of similar tubes, which arise from all
parts of the body, and are termed *absorbents.*
The whole of the absorbing tubes, after passing
through various glands, at length unite into one
or two of larger size ; the recipient tube on the
left side being by far the largest, and known by
the appellation of the *thoracic duct.* These
larger absorbent tubes pour their contained
fluids into the veins named the *sub-clavian;* and

thus into the general mass of the blood. The
exact nature of the changes which the chyle
and the lymph undergo in their passage through
these tubes, is not well understood. One change
appears to be, that the chyle, as first formed in
the alimentary canal, is to a certain extent, *completed*, or freed from water, during its course
through the lacteals : for though, when the
chyle is mixed with the blood, its albuminous
principles are much less perfectly developed
than the principles of the blood itself; yet the
developement of the albuminous principles, on
their mixture with the blood, is more perfect,
than when the chyle is first taken up from the
alimentary canal.

The matters conveyed from the other parts of
the body, by the tubes of the general absorbent
system, have, by most physiologists, been supposed to be of an excrementitious character.
That some of the absorbed matters are excrementitious, is very probable : arguments may,
however, be adduced, to show, that the whole of
the matters absorbed are by no means excrementitious ; but that they are repeatedly con-
signed to the uses of the vital agency : every
new organization raising them, as it were, a step
higher, and qualifying them for those refined and
ulterior purposes; for which the crude chyle can
hardly be imagined to be at once adapted.

The circumstances favouring the above opinion, which we are now desirous to mention, are,—

First. It is unreasonable, and contrary to every thing we know respecting the operations of the animal economy, to suppose that the chyle should be separated from one kind of excrementitious matter, in the alimentary canal ; in order to be immediately mixed again with other excrementitious matters, in the chyliferous tubes. It is, therefore, a just inference, that .if the matters contained in the absorbents, were really and wholly excrementitious, they would be carefully kept apart ; and would be removed from the system by some other means, than by tubes united with those conveying the nutritious fluids.

Secondly. By admitting that the fluids contained in the absorbent tubes possess a highly animalized character ; the design of their union with the crude and imperfectly animalized chyle, becomes apparent : the fluid in the absorbents will be seen to execute an important and necessary office ; by raising the vital character of the chyle, and qualifying it, for becoming a part of the general mass of the blood. We thus obtain a cogent reason, why the fluids taken up from the internal surface of the alimentary canal, should be mingled with the fluids absorbed from the other parts of the body ; a mixture which is inexplicable, on the hypothesis of these absorbed fluids being wholly excrementitious.

Thirdly. The gradual developement of the staminal principles of animal bodies, by repeated organizing processes, fully accords with those general views of the operations of nature, which, throughout this work, we have endeavoured to illustrate; and which lead to the general conclusion, that the operations of nature are never abrupt, but always slow and gradual. Further, it is more reasonable to conceive, that matters already assimilated to the animal body, are better fitted for its immediate uses; than matters which, like the chyle, have only received an imperfect assimilation.

Fourthly. Many animals can and do live, for a considerable time, on substances contained in their own bodies. Thus, hybernating animals, as previously stated, have the ability to assimilate further, those matters which have already become part of themselves; consequently, such a faculty of progressive organization as we have supposed, actually exists; and *a sort of digestion is carried on in all parts of the body, to fit for absorption and future appropriation, those matters which have been already assimilated.* Were it necessary, other arguments to the same effect might be added : but we shall at present delay the further consideration of the assimilating character of the whole absorbent system; that we may recur to it again, in a succeeding part of the present chapter.

2. *Of the Blood.*—The blood is that well-known fluid pervading the tubes, named from their function the *blood vessels;* which tubes are extended more or less over every part of an animal. We have already described the general distribution of the blood vessels; and shall now confine ourselves chiefly to the properties of the blood itself.

The chyle, as we have stated, is poured into the general mass of the blood near the heart; and from the heart is almost immediately propelled through the lungs. The chyle, thus set in motion, is not only united thoroughly with the blood; but undergoes those other important changes, by which its final conversion into blood is accomplished. The exact nature of these changes is unknown; but they are evidently of a completing character—that is to say, the weak hydrated ingredients of the chyle, are freed from a portion of the water with which they were associated; and are transmuted into the strong albuminous matter of the blood.

The chief constituents of the blood are essentially albuminous. Blood contains albumen in three states of modification: namely, *albumen,* properly so called; *fibrin;* and the *red particles.* In addition, there are *oily* matters; besides various minute portions of other animal matters, and *saline* matters, all dissolved, or rather suspended, in a large quantity of water. The following short table exhibits the relative proportions of the constituents of human blood to each other, as they exist in most individuals.

ONE THOUSAND PARTS OF HUMAN BLOOD CONTAIN

Of Water	783,37
Fibrin	2,83
Albumen	67,25
Colouring matters	126,31
Fatty matters, in various states	5,16
Various undefined animal matters, and salts .	15,08
	1000,00*

The reader will not fail to remark, that among these constituent principles of the blood, gelatine is not mentioned. In fact, though existing most abundantly in various animal structures, *gelatine is never found in the blood, or in any product of glandular secretion.* We formerly noticed that in the scale of organized substances, gelatine appears to rank lower than albumen : and we may now add, that a given weight of gelatine, contains at least three or four per cent. less carbon, than an equal weight of albumen. The production of gelatine from albumen must, therefore, be a *reducing* process. We shall presently have occasion to revert to these facts. In the mean time we subjoin the few observations we have to offer, on the organization or structure of the blood.

* Le Canu ; mean of two analyses.

The organization of the blood is even more wonderful than its chemical composition, and is still less understood. The red portion of the blood, for example, is composed of innumerable minute globules, varying in size in different animals; and in all instances, highly organized: the real structure indeed of these globules is very imperfectly known; but they are generally supposed to be formed of solid colourless nuclei, within red vesicles. The fibrin, also, is diffused through the mass of the blood in a state of equally minute subdivision; though the particles of the fibrin are colourless, and their magnitude much less than the magnitude of the red particles. From this inferiority in size, some physiologists have been led to think, that the colourless particles of the fibrin, are identical with the nuclei of the red particles. During the life of an animal, the particles of the fibrin, as well as the red particles of its blood, seem to be in a state of extreme self-repulsion; by which self-repulsion, the union of these particles is prevented; except as the economy of the animal may require, and may determine. After death, however; or in blood withdrawn from the body of a living animal, the property of self-repulsion, more especially among the fibrinous particles of the blood, ceases, and they readily cohere: this cohesion is termed the *coagulation* of the blood. Much beautiful design is probably concealed under that peculiar organization of the blood, to which it owes its coagulating tendency. One result of the coagulation of the blood, indeed, is as obvious as it is important; namely, the prevention of hæmorrhage. If the blood did not coagulate, the existence of animals would be most precarious; as on the slightest injury, they would be liable to bleed to death.

3. *Of Respiration.*—The function of *Respiration*, or breathing, is, perhaps the most important in the animal economy: many of the other functions may be suspended; but the interruption of breathing is immediately destructive of life. When we described the phenomena of the circulation of the blood; we observed, that the blood, in passing through the lungs, is exposed to the action of the atmospheric air. Now, during this exposure of the blood to the atmospheric air, it undergoes certain changes. The blood from the right side of the heart, when it enters the lungs, is of a dark red colour: the blood is then dispersed, in a state of most minute subdivision, through the ultimate vessels of the lungs; and in these vessels is brought into contact with the atmospheric air, when it becomes of a bright red colour. In other words, the blood changes in the lungs its *venous* appearance, and assumes the character of *arterial* blood. The blood thus arterialized, returns to the left side of

the heart; and from the left side of the heart is propelled through the whole arteries of the body. In the minute terminations of the arteries, the blood again loses its florid hue, and, reassuming its dark red colour, is returned through the veins, to the right side of the heart; to be exposed as before to the influence of the atmospheric air, and to undergo the same succession of changes.

On examining the respired air, a remarkable alteration of its properties is found to have taken place; a portion of its oxygen has disappeared, and a similar bulk of carbonic acid gas, has been substituted. With respect to the origin of this carbonic acid gas, there have been various opinions. Formerly, the greater number of physiologists maintained, that carbon, in some form, was excreted by the lungs; and that this excreted carbon, uniting with the oxygen of the inspired air, was converted into carbonic acid gas. No one imagined that oxygen gas could be passing inwards through the membrane of the lungs; while carbonic acid gas was, at the same time, passing outwards, through the same membrane. Accurate observations have, however, demonstrated, that such a simultaneous passage of gases really takes place through the membrane of the lungs: and the observations are not confined to the two gaseous bodies in the lungs; but are applicable to all gases what-

ever, under similar circumstances. In consequence of these observations, it seems now to
be generally admitted, that the oxygen of the
atmospheric air is absorbed by the blood ; and, in
some unknown state of combination, reaches the
extreme subdivisions of the arteries ; where it is
united with a portion of carbon, and forms carbonic acid gas : that this carbonic acid gas is
retained in some unknown state of combination
in the venous blood ; till, in the lungs, it is
expelled, and oxygen is absorbed in its stead ;
according to the laws which regulate the diffusion of gaseous bodies, and which were formerly explained. Along with the carbonic acid
gas, a large quantity of aqueous vapour, as we
have stated, is at the same time separated from
the blood.

It would be foreign to the objects of this treatise, were we to enter further into the reasons
for the view we have given of the phenomena
of respiration. These reasons are many and
strong ; and seem indeed to prove clearly, that
the changes which the blood undergoes, during
its circulation through the body, are as we have
described them. We shall, therefore, assume,
that our view of respiration is correct ; and shall
offer a few remarks on the attendant circumstances, and on the consequences, of respiration.

First. To what influence are we to ascribe the
different colours of arterial and of venous blood?

The opinion formerly held, was, that the arterial colour arose from the absorption of oxygen; and the venous colour from the presence of carbon. But recent observations seem to show that the change in the colour of the blood during its circulation, if not entirely independent of oxygen, is much influenced by the saline matters; particularly by the common salt, which the blood contains: and that the dark colour of venous blood, is principally owing to the presence of carbonic acid gas.

Secondly. What is the source, of the carbonic acid in venous blood, and of the aqueous vapour which is expelled from the lungs? These questions cannot be answered with certainty. But some observations lately made, have induced us to believe, that the *conversion of albuminous matters into gelatine,* is one great source of the carbonic acid in venous blood. Gelatine, as before observed, contains three or four per cent. less of carbon than albumen contains. Now gelatine enters into the structure of every part of the animal frame, and especially of the skin: the skin indeed consists of little else besides gelatine: it is most probable, therefore, that a large part of the carbonic acid of venous blood is formed in the skin, and in the analogous textures. Indeed, we know that the skin of many animals gives off carbonic acid, and absorbs oxygen; in other words, performs all the offices of the lungs:

a function of the skin perfectly intelligible, on
the supposition that near the surface of the
body, the albuminous portions of the blood are
always converted into gelatine. With respect to
the aqueous vapour thrown off from the lungs :
we have every reason to believe, as before stated,
that much of this vapour is *derived from the
chyle*, in its passage through the lungs ; and that
by such separation of water, the *weak* and delicate albumen of the chyle, is converted into the
strong and perfect albumen of the blood ; according to the principles detailed at the commencement of this chapter.

Thirdly. What are the uses of the continual
extrication of carbonic acid from living animals ;
and could not a little superfluous carbon have
been thrown off from their bodies in a more
simple manner ? The precise use of the constant
evolution of carbonic acid, or how it is effected,
we know not ; but one great use which has been
assigned to this evolution, is, the formation of
the heat of the body ; and not only the power of
forming that heat ; but also the power of varying
it according to circumstances—a power so characteristic of organic life. Out of the body, carbon does certainly give off heat on combination
with oxygen. Hence, it has been maintained
with great plausibility, that the same combination, within a living body, may give origin to
its heat ; though it must be confessed, there are

some difficulties about this view of the origin of animal heat, which detract considerably from its likelihood. Moreover, it is exceedingly probable, that though the evolution of carbonic acid gas, may be one of the means possessed by the animal economy for generating heat; there are yet other means, the nature of which at present is quite unknown.

The quantity of carbon thrown off by the lungs, is very abundant; but has probably been much overrated. Philosophers have, for instance, calculated that the lungs of a man of ordinary size expel, in the course of twenty-four hours, eleven ounces of carbon : a quantity of carbon more than equal to the quantity contained in six pounds of beef.* If carbon be indeed thrown off from the lungs so copiously ; it must be produced within the body. It is difficult to account for the quantity of carbon thrown off, even on the lowest estimate. We are, therefore, necessarily obliged to conclude, that more solid matter is every day expelled

---

* According to an elaborate analysis, by Berzelius; the muscle of an animal contains 77 per cent. of water, and 23 per cent. of other matters. Supposing, what is near the truth, that 22 of these 23 parts consist of albumen, and that this albumen contains half its weight of carbon ; which in round numbers is a sufficiently near approximation ; it follows, that 100 parts of the muscular fibre of animals, contain about 11 parts of carbon; so that 11 ounces of carbon must represent 100 ounces of beef; which is upwards of six pounds, as stated in the text.

from the body by the lungs, than in any other manner. Hence the probability of the opinion formerly noticed, that the matters taken up by the absorbents, and by the veins, enter successively into the formation of various parts of the animal frame ; instead of being removed, immediately after their absorption, as at present is generally supposed. For it seems hardly possible, to reconcile, with the quantity of food, the great quantity of carbon that is expelled from the lungs alone ; much less, what must be expelled, if all the matter taken up by the absorbents be likewise considered excrementitious.

4. *Of Secretion.*—From the blood, are formed, by means of peculiar apparatus, all those numerous products termed *Secretions ;* not only so unlike each other, but so unlike the fluid from which they are separated. Some of these secreted products appear to be little else, than a disengagement of certain matters already existing in the blood. Other secretions have no resemblance to any ingredient of the blood : consequently, in the glandular structure, by which these secretions, so dissimilar to the blood, are formed, the blood must undergo some essential change. In the present state of our information, however, we must content ourselves with a limited insight into the nature and the causes of secretory action. We see that se-

creted products are of two kinds : that some of
the matters separated by animal bodies are
evidently thrown off, on account of their noxious
qualities; are, in fact, *excretions:* which could
not be retained without proving fatal to the life
of the animal from which they are detached :
while other matters separated from animal
bodies, are as evidently intended for further objects, and for the performance of various subordinate actions in the living system ; are in
fact, *secretions;* properly so called.   But as we
have stated, we are still perfectly unacquainted
with the intimate nature of the changes which
produce the secreted fluids; though it is probable that a careful examination of the phenomena, would throw much light on the general
character of these changes ; and would display
evidence of the most consummate design.

5. *Spontaneous Decay of Organized Bodies.*—
It remains finally to close this work with a few
observations on the spontaneous, and inevitable
decay that awaits all the things produced by
organization; after they have been removed
from the influence of those organic agents, by
which the combination of their constituent principles was effected.

The organized beings that inhabit this globe,
however numerous, bear a very small relation to
the magnitude of the globe, and seem to occupy

its surface only. We have seen that the elements forming the structure of organized beings,
are not only combined in different proportions;
but that in many instances, these elements appear to undergo further decomposition into ultimate forms of matter; which, out of a living
body do not, and perhaps, in the present constitution of the universe, cannot, exist in an
isolated state. Owing to this diversity in the
composition of organized beings from the composition of inorganic matter; and to other causes
which will readily occur to the reader, organized
beings and their laws, are in continual opposition to the general laws, by which inorganic
matter is governed. To counteract, therefore,
these opposite laws, and to maintain the existence of organized beings, demands the unremitting efforts of the organic agency. But at
length these efforts are exhausted; the contest
ceases: when the general laws of inorganic
nature prevail; and speedily reduce, to their
original state of existence, the atoms which had
been incarcerated in the living frame.

The spontaneous decay of organized beings
is usually termed the *putrefactive* process; and
some substances have much more tendency than
others, to undergo putrefaction. As might be
expected; substances whose constitution is most
simple, as the oils, and bodies of a like nature,
are also the most permanent; while substances

more compounded, especially those substances
which include azote, are exceedingly liable to
putrescent change. For such changes, a certain
degree of heat and of moisture appear to be
necessary: since in a temperature below the
freezing point of water, or in a perfectly dry
state of the atmosphere, even animal substances
may be preserved unchanged, during any length
of time. The phenomena resulting from the
dissolution of different kinds of organized matters are of course different; but in every instance, the tendency is toward the formation of
compounds more simple than the matter decomposed; that is to say, of compounds whose existence, out of a living body, is not incompatible
with the present state of the globe. The matters which, in a warm and damp air, seem first
to be loosened from organic combination, are
those foreign bodies we have already mentioned, as existing, in every part of the structure
of organized beings, in some unknown but active
self-repulsive state. Hence arises, during putrescent changes, the formation of sulfuretted and
phosphoretted hydrogen, and of other undefined
compounds of the same elements: and these
gaseous compounds, chiefly, produce the very
offensive odour of putrefaction. At the same
time, there are formed, carburetted hydrogen,
oil, acetic acid, ammonia, and last of all, carbonic acid gas and water; while the azote is

extricated in a gaseous condition. Finally, both vegetable and animal matters, but vegetable matters more especially, are reduced to the state of mould. The mould from vegetable matters, consists principally of carbon, in combination with a little oxygen or hydrogen : the mould from animal bodies, contains the same elements as vegetable mould, together with a little azote, and the usual saline ingredients of organized substances. In this form of mould, the remains of vegetables and of animals, as was before stated, constitute the food of plants. By plants these remains are again organized, and thus go through the same series of changes.

We may here pause for a moment, and, on account of the general reader, briefly recapitulate the most striking facts, which have been detailed in the present, and in the preceding chapters.

In the first place, the *mechanical arrangements* for reducing the food of animals to the proper degree of comminution, are wonderfully varied ; according to the peculiar qualities of that food. In the graminivorous and granivorous tribes, for example, the teeth are literally instruments for grinding or triturating herbaceous matters, and seeds. In carnivorous animals, such a structure would be useless : the teeth, therefore, are suited only for cutting, or tearing. In gnawing animals, the teeth present a totally different structure, but at the same time are admirably fitted to the habits of the animal. Occasionally, as in the fowl tribe of birds, the grinding apparatus is placed, not in the mouth, but in the stomach itself; this organ being, as it were, expressly contrived for trituration; while some of the functions the stomach performs in other animals, are transferred to contiguous parts.

The structure and mechanism of the stomach, and of the alimentary canal, then claim our particular attention. In carnivorous animals, whose food requires comparatively little assimilation, the alimentary canal is short, and of a simple structure. On the other hand, in vegetable feeders, that canal is long and complicated; but perfectly adapted for macerating their food, and for extracting from it, every thing that can be converted into nourishment. Nor is there an adherence to any model, but the whole alimentary structure is throughout varied; as if in order to demonstrate the power and the wisdom of Him, by whom the organization of animals has been contrived. Thus the alimentary canals of the cow, and of the horse, are formed on entirely different models; though the food of both animals is nearly the same.

We proceed in the next place, to the consideration of the *chemical changes*, which the food undergoes in the stomach and duodenum. In these changes we discover arrangements not

less wonderful, indeed more wonderful, than in
the arrangements of structure and of mechanism.
The variety of forms, assumed by bodies having
the same essential composition, produces a latitude, in the choice of diet, which is almost
infinite : at the same time, the organs being
endowed with the power to discriminate all these
differences, and to act on the ultimate principles
of bodies, elaborate the same uniform chyle,
from every variety of food. The power by
which the stomach is enabled to effect these
astonishing changes, is the power which it possesses, of associating the different alimentary
substances with water; the power, in short, of
dissolving, or digesting them. This dissolving
power seems to be exerted by the stomach,
through the agency of chlorine derived from the
common salt in the blood ; at least, chlorine is
always present in the stomach, during the act
of the solution of the food ; though the precise
mode in which the chlorine operates, is still unknown. Contemporaneously with the act of
solution of the food, such essential changes take
place in its composition, as are requisite for per-
fecting the future chyle.

The stomach having accomplished its office,
the digested mass enters the duodenum ; where
the series of changes is continued in a manner
equally wonderful. In the duodenum, the digested mass is brought in contact with the

biliary and the pancreatic fluids. The alkali of the bile unites with the acid, with which the food had been mingled during its digestion in the stomach ; the excrementitious parts, both of the food, and of the bile, are separated or precipitated ; while at the same time, the proper chylous principles are eliminated, in a condition appropriate for their absorption by the lacteals.

There are two divisions of those minute tubes, composing what is termed the absorbent system of animals ;—the lacteals—and the absorbents properly so called. The ultimate ramifications of the lacteals, originate from the internal surface of the alimentary canal, where they take up the digested, and partly assimilated, aliment, or chyle. The ultimate ramifications of the proper absorbents, originate from all parts of the body ; and are enabled to take up, by some peculiar process, every component of the body, solid as well as fluid, in the same manner as the chyle is taken up by the lacteals.

The fluid obtained from the lacteals, and the fluid obtained from the proper absorbents, are both alike albuminous. The albumen of the chyle, as we have formerly shown, is produced in the stomach and duodenum, while the food is undergoing the process of digestion. But whence is derived the albumen, found in the proper absorbents? The animal body we know to be composed of a great variety of matters, among which gelatine predominates. Now, since albumen only, is found in the absorbents; it follows, that the gelatine of the body previously to its being taken up by the absorbents, is re-converted into albumen : in other words, the absorbed gelatine undergoes a process, entirely analogous to that process which gelatine, and other matters, undergo in the stomach and duodenum, during the process of digestion. Hence, the digestive process, instead of being confined to the stomach and duodenum, is actually carried on without intermission, in all parts of a living animal body.

The two kinds of fluid albumen derived from these two sources ; that is to say, the crude chyle in the lacteals, and the highly animalized lymph in the absorbents, are at length commingled ; and form one uniform fluid of an intermediate character, adapted for becoming a part of the general mass of the blood. The character, however, of this fluid, when it becomes part of the blood, though albuminous, is still very *weak;* that is to say, the commingled fluid from the lacteals and from the absorbents consists of albumen, holding a large proportion of water in a state of essential combination. By a beautiful arrangement, as soon as this weak albuminous fluid has joined the stream of blood, it is hurried through the lungs ; where it undergoes a remarkable change. In the lungs, the water, which we have shown to be in essential union with the weak albuminous matter of the chyle, is separated from that state of union, and is expelled along with the carbonic acid gas, which is continually escaping from the lungs: the weak and delicate albuminous matter of the chyle, is thus converted into the strong and firm albuminous matter of the blood. We are next brought to consider the process of respiration.

The blood, in its course through the lungs, emits carbonic acid gas, and assumes a florid *arterial* colour. In the lungs, the blood also, according to the principles of gaseous diffusion, absorbs a portion of oxygen from the air of the atmosphere. The oxygen thus absorbed, remains in some peculiar state of union with the blood, (Query, as oxygenated water, or some analogous compound?) till the blood reaches the ultimate terminations of the arteries. In these minute tubes the oxygen changes its mode of union: the oxygen is converted into carbonic acid by combining with a portion of carbon derived from the albuminous principles of the blood; at the same time, heat is extricated. Two distinct alterations take place during the union of the carbon with the oxygen in the ultimate terminations of the arteries: First, a portion of the albumen contained in the blood, is supposed to be reduced to the state of gelatine; which gelatine is appropriated toward the renovation of those textures whose composition is chiefly gelatinous. Secondly the carbonic acid formed from the reduced albumen, unites with the blood ; communicates to the blood its dark *venous* colour ; and is transferred to the lungs. By the lungs the carbonic acid is expelled from the system ; along with a portion of aqueous vapour, derived principally from the weak albumen of the chyle ; as formerly explained.

The blood is the source, not only of all the constituent principles of animal bodies, but likewise of all the various *secretions ;* many of which secretions differ altogether, in their properties, from those of the primary fluids ; and perform secondary offices, of great importance in the animal economy.   Other products separated from the blood, are purely *excretions ;* as, for instance, the carbonic acid gas from the lungs ; which could not be retained in the animal system without destroying life.

Finally, the life of the animal becoming extinct, the essential properties of the matter of which it is composed, resume their natural action, and speedily restore the elements to their original condition.

Such is a summary of the operations of living bodies, which, in the present and in the preceding chapters, we have endeavoured to illustrate.   Our insight into those operations, though very imperfect, is amply sufficient to satisfy us, of the infinite wisdom by which they are directed ; and that the unknown, must be far more wonderful than what has been disclosed. Most of the facts, also, on which we have dwelt, are of a character so obvious, that they require only to be understood, in order to be admitted among the proofs of the great argument of design ; at least, by all, but persons who affect to deny that argument. We therefore leave to the reader, the application of facts, so obviously demonstrative of design; and proceed to offer a few remarks on certain general arrangements of organized and living beings, relatively to the arrangements of inorganic matter.

First. In considering the economy of organized beings, one of the circumstances most calculated to arrest our attention, is the extraordinary skill manifested in the disposal of the various parts of the organized system, with regard to each other. As an instance, on the great scale, may be noticed, the mutual relation and dependance of plants and animals. Thus, as we formerly pointed out, carbonic acid gas constitutes the chief food of plants; and we now see, that nearly the whole of the superfluous carbon produced by the operations within animal bodies, is actually thrown off in the form of carbonic acid. Plants, therefore, on the one hand, supply the chief nourishment to animals; while that gaseous matter which is separated by the animal economy, and which if retained within animals, would to them be fatal, constitutes, on the other hand, the chief food of plants. Nor in these two respects only, are the two great systems of organization mutually dependent: for unless plants consumed the carbonic acid gas which is formed by animals; this deleterious compound would probably accumulate in the atmosphere, so as to destroy animal life; while it is doubtful, whether the present races of vegetables could exist, if carbonic acid gas were not formed by animals. Again, the general scheme of Providence, for the nourishment of animals, claims our especial notice. Animals have not only been destined to prey on each other: but all created beings are the food of other beings progressively higher than themselves in the scale of organization. By this wise arrangement, the labour of the assimilating power has been greatly diminished; and by the same means, that accumulation of dead animal remains, which soon would be overwhelming, is entirely prevented. Even in the fabric of individual animals, and in the operations which are carried on within these beings, the same wise purposes of mutual relation and dependance are observable. Thus, whether we contemplate the repeated employment of the same materials; or the various important ends, in many instances accomplished by the same process ; we discover, throughout, the utmost abridgement of labour ; so that the greatest possible effect, is every where produced, by the simplest possible means.

Secondly. The general subserviency, of the mechanical arrangements of the frame of organized beings, to the chemical operations that are carried on within them, is of still greater interest and importance even, than the mechanical arrangements of the frame have been shown to be. We may view an organized being as a piece of intricate machinery, adapted to the physical, and the chemical properties of matter. The adaptation of this machinery to the physical properties of matter, belongs to another department. Our attention is directed solely to the chemical adaptations. The performance of the chemical changes within organized beings, through the interposition of mechanical arrangements, as has been stated in a former part of this work, establishes, beyond a doubt, that these chemical changes have a real existence. Thus, when we witness such a display of elaborate arrangements, as are exhibited in the mechanism of the digestive organs, and of the circulating system ; the purpose of which arrangements is merely to produce a few chemical changes in the food, and in the blood ; it is evident that the chemical changes so produced, must be at least as real,

as the mechanical structure, by means of which
they are effected. The adaptations of mechanical arrangements in the structure of organized beings, to the pre-existing chemical properties of matter, afford also an evidence of
design, not less impressive than unequivocal.
The most determined sceptic cannot assert that
there is any *necessary* relation, or indeed any
relation whatever, between the mechanical arrangements, and the chemical properties to
which they administer.   There is no reason
why the chemical changes of organization,
should result from the mechanical arrangements, by which they are accomplished : neither
is there the slightest reason, why the mechanical
arrangements in the formation of organized
beings, should lead to the chemical changes of
which they are the instruments.   From what
cause, then, arose the association of the chemical
changes with the mechanical arrangements?
How were the chemical operations of digestion
and of respiration brought into union with the
mechanical apparatus of digestion, and with the
circulating system?   The co-existence of things
so entirely dissimilar, and having no kind of
mutual relation, can be explained only on the
supposition that *a will exists somewhere;* and also
a *power to execute that will.*   The existence is
thus unavoidably acknowledged of a Being,
who knowing every pre-existing chemical pro-

perty of matter, and willing to direct these chemical properties to a specific object, has contrived for that purpose an apparatus admirably fitted to attain His object. Such is the explanation—the only possible explanation, of the subserviency of mechanism to chemistry, in the processes of organic life. And what is this explanation, but our *argument of design*, in terms that seem absolutely irresistible?

Thirdly. The perpetual renovation and decay to which all organized beings are liable, may be considered as a part only of the great round of changes we witness in every thing that has been created. The world itself, as we have seen, appears to have been, at intervals, subjected to changes involving even the fundamental laws by which it is governed. Nothing, therefore, belonging to the world, can reasonably be expected to be permanent. Had there been even an approach to such permanence, the beautiful adaptations of organized beings to the pre-established laws of inanimate matter, and all the other wonderful arrangements we have described, could not have been manifested as they now are. Besides, to the changes we ourselves undergo, we are indebted for the greater part of the enjoyments of our life. If none died, none could be born; and the present arrangements of human society could have no existence. There would be none of the pleasing

relations of parent and offspring; none of the
agreeable variety of childhood, of youth, of
maturity, and of age, experienced by every
individual; which, with all the other numerous
relations of society, incidental to the persons of
different individuals, contribute so largely to
human happiness. Were man exempt from
change; whether the rest of the world, were
supposed to be progressive, as it is; or to be
stationary; as regards him, the same uniform
and dull monotony would prevail, the same want
of motive. In short, with our present constitution and feelings, perpetuity and uniformity are
physically and morally impossible.

But why, it has a thousand times been asked,
why has the world been so constituted? Why
this unceasing round of change? Whence its
origin? What its object?—Such questions, the
Great Author of the Universe alone can answer.
But as within those narrow limits by which our
observations are bounded, wherever we can trace
His designs, we see that His works are never
without an object; we cannot doubt that in determining their perpetual change, there is no
less an object; though the object be above our
comprehension. By placing immaterial and intelligent beings, for a time, in personal connection with matter, He has indeed communicated
to them a knowledge of those properties of
matter which so strikingly display His wisdom

and power ; and this may have been one of His objects :—but to speculate further on points so utterly beyond our capacity, would be presumptuous : for who can " know the mind of God, or who hath been His counsellor ? "

We have thus given a brief outline of what has been denominated the Chemistry of organization ; in other words, an account of those changes and combinations which, through the operation and the agencies of inorganic matter, organic agents are capable of effecting. The information it has been in our power to give, though imperfect, we have shown to be amply sufficient, not only to demonstrate the astonishing wisdom, and foresight, with which organized beings, in as far as we can understand them, have been contrived and formed ; but the infinitely higher perfection of both these attributes, requisite to impart to organization, that vitality, the nature of which so entirely surpasses our conception.

We shall now close this volume with a few observations on the future progress of chemistry ; on the means by which this science may be applied to physiological research; and on the tendency of physical knowledge in general.

Chemistry, as we pointed out in the introduction to this treatise, forms the connecting link between that kind of knowledge which is founded on quantity; and those kinds of knowledge which rest solely on experience. All our experimental knowledge that is not chemical; for instance, all physiology relative to the phenomena of life, is wholly removed from the logic of quantity, and depends entirely on observation. Now so far as the logic of quantity is applicable; so far are we certain of our conclusions; as certain at least as we are of our own existence, or that we see and hear. But when this logic cannot be applied; our conclusions are no longer such as *must* be—no longer follow from our premises as a necessary consequence; but are only, for the most part, such as *may* be; that is to say, have no more than that degree of probability, which arises from the evidence we have of the truth of the phenomena or events, forming our premises.

In all knowledge depending on mere observation or experiment; what we know, is grounded on our own observation and experience, or on the observation and experience of others. What we ourselves observe, we too often observe very imperfectly; or do not understand, when observed. But phenomena or events, the knowledge of which we are obliged to receive at second hand, on the *testimony* of others; and which may have been observed through the distorted medium of ignorance or of prejudice—may even have been wilfully misrepresented—of these phenomena we have a still less assurance. If a phenomenon or event has happened only once, and be therefore historical; we are under the necessity of acquiescing in its truth, or of estimating its probability, according to the rules of evidence. If the phenomenon or event be of frequent occurrence, or if it be capable of being brought under our own observation; in order to remove our uncertainty, we endeavour to observe it ourselves; in short, we make an experiment. Such is the method we pursue, in obtaining all that knowledge which is the result of mere observation. The different events succeed one another, but we know not wherefore; we see not their mutual connexion. We believe that an event will, *probably*, follow another event; because the one event has always followed the other; or because of some other probability: but we cannot discover that *necessary* connexion between the two events, which so irresistibly leads us to determinate conclusions, where we can apply the laws of quantity.

The foregoing observations may be viewed as a continuation of those which were offered as preliminary to the first treatise in this volume;

and relate chiefly to the progress of chemistry. Chemistry being a science of observation, we can form but a very imperfect conception of its future progress; because, we cannot, by reasoning, anticipate the discovery of those chemical facts which are yet concealed. The progress chemistry has made within these few years, is truly astonishing; and when a more rigorous mode of observation shall be adopted—in short, when chemistry shall be brought more under the control of the laws of quantity—a control which will be exercised, at least indirectly— it is impossible to foretell the degree of perfection, chemistry may attain as a science. But, for many years yet to come, the progress of chemistry must depend solely on experiment; and its cultivators must be satisfied with the comparatively humble office, of discovering the actual chemical changes, which bodies effect on each other.

Since, then, in knowledge derived from observation, an acquaintance with *what exists*, and with *what is done*, is indispensable: to obtain a clear, accurate, and unequivocal conception of natural objects, and of the changes to which they are liable, is the first duty of every observer, and of every experimentalist. Nor is there any observer, or experimentalist, however unpretending, who may not add to the stock of ascertained facts; so varied and inexhaustible are the stores of nature. Another duty of every one who engages in observation or experiment, is to become the faithful historian of what he witnesses ; to narrate in plain and intelligible language, the event or phenomenon he has observed ; so that a just notion of it may be conveyed to others. We say a *just* notion ; in the greater number of instances, a *perfect* notion is impossible ; because what is seen, cannot be expressed by language. But to give a just notion ; that is, a notion which, though incomplete, *has no foreign or false gloss;* is within the power of every observer ; and to give such a notion of the facts he narrates, ought to be his chief study. One testimony of so faithful a witness is often invaluable, and worth a thousand vague and inaccurate observations ; which are only calculated to bewilder, or to mislead ; and thus are worse than useless.

The next rule which an interpreter of nature ought to bear in mind, is *not to attempt too much at first;* but in order to establish a sure foundation for his succeeding labours, be content with obvious and unexceptionable facts ; and so arrange these facts, that they may lead to others. To elicit novel and prominent facts, is the lot of few ; and any one may happen to be so successful. But all, as before stated, may *investigate truth;* and thus contribute more or less towards the advancement of knowledge.

Moreover the humblest contributors may rest assured, that they are imperceptibly raising a structure, which will sooner or later include the conspicuous labours of their more fortunate co-adjutors; in which structure, these labours will indeed still appear conspicuous; though their importance will be diminished as the fabric is extended around them.

The remarks just made, have especial reference to the application of chemistry to physiology. The cautious and judicious application of chemistry to physiology has already effected much, and is capable of effecting still more. Indeed, no one can pretend to be a Physiologist who is ignorant of chemistry. But chemistry, in its present state, in order to be made really useful, must be applied in a manner the most guarded and sparing—must, indeed, be rigidly confined to the ascertainment of *what the living principle does;* and *how it operates on inorganic principles.* With the living, the animative properties of organized bodies, chemistry has not the smallest alliance; and probably will never, in any degree, elucidate these properties. The phenomena of life, are not, even remotely, analogous to any thing we know in chemistry, as exhibited among inorganic agents. The great error of chemists, therefore, has been their attempting to apply that science to explain phenomena, for the explanation of which, chemistry, as we have said, is totally valueless. Such perversion of the reasoning powers, has too much prevailed among physiologists in all ages. In the earlier ages, heat was considered the principle of life. In later times, electricity has been discovered; and to electricity, the same functions have been ascribed. Life, according to other philosophers, is motion. But the progress of science has dispelled all these illusions: the origin of the obscure and evanescent principle of life, must be sought.elsewhere. By heat, for example, many wonderful things may be accomplished; but heat will not act of itself. The powers of electricity are still more wonderful than the powers of heat: but electricity, we know to be governed, in its mode of action, by certain laws; and we know that electricity gives no sign of intelligence. In the same manner, life, as we are acquainted with it, cannot exist without motion; but motion can exist without life. Life and motion, consequently, are not synonymous terms; nor can we conceive the existence of motion, without a mover. In short, the living principle, as we have already shown, is something different from, and superadded to the common agencies of matter; over which, to a certain extent, it has a control. Thus, the phenomena exhibited by the mysterious agency of life, are strictly comparable only with one another; and have no relation to any inorganic phenomena.

But the desire of the Physiologist to ascribe to the agencies of inorganic matter, those operations ascertained to be carried on within living bodies, is merely a display of that innate propensity of the human mind, by which we are led to seek after First Causes. The conceptions of the physiologist regarding the principle of life are the same, therefore, as the conceptions of mankind in all ages regarding the Great First Cause—the Deity himself. The poor untutored savage " sees God in every cloud, and hears him in the wind." The complacent philosopher smiles at the credulity of the savage, and perhaps deifies " the laws of nature!" Both are alike ignorant; nor is the imagined Supreme Being of the untaught savage, in any degree, more absurd, than the imagined Pantheism of the philosopher. The winds we know can be referred to other causes, to which they are immediately owing: so with the progress of knowledge, the " laws of nature," have been found to merge, and will continue to be found to merge, into other laws, still more general; thus proving that these " laws of nature" are, all alike, mere delegated agencies.

Hence the tendency of knowledge, and of its due application, is to abstract the attention from inferior things, and to fix the mind on the source of all knowledge and of all power— the GREAT FIRST CAUSE; who exists and acts throughout the universe; whom we can approach only, by studying His works; and whose works, an eternity, will be inadequate to explore.

APPENDIX.

Page 59.—*Elementary form of electrical energies, &c.*
Throughout this work, we have adhered as much as possible
to the common language of chemistry. We conceive, however,
that the phenomena of chemistry may be expressed in terms
of hypotheses, of which the chief are the following :—

1. That every portion of matter *attracts,* and *is attracted* by,
every other portion of matter, according to laws which have obtained universal assent.

2. That all matter, as it is known to us, exists in the condition
of *molecules;* which molecules we consider to be virtually
spheres or spheroids.

3. That all the spherical or spheroidal molecules, when unim-
peded, have a tendency to *revolve on their axes,* with velocities,
which in molecules having the same weight, are, under similar
circumstances, fixed and definite ; but which velocities, according to a law which need not be here specified, increase in molecules of different weights, as the weights of the molecules
diminish.

4. That the *polarity* of molecules results from the motion
upon their axes ; that fluids exhibit no *sensible* polarity, because
the motions of the contiguous or alternate molecules of which
fluids consist are opposite and equal, and thus exactly neutralize
each other's effects ; that polarity becomes *sensible* in fluids,
when molecules are separated from the neutralizing influence of
the contiguous molecules, and *move together alone in the same
direction;* that the *opposite* motions of contiguous molecules
thus separated and moving together alone, are the cause of
opposite (positive and negative) polarities ; that the *intensity* of
the polarity of molecules is of two kinds, the intensity depending
on the greater or less velocity of the molecules, and the in-

tensity depending on the greater or less separation of contiguous molecules ; finally, that the *quantity* of polarity is proportional to the number of molecules moving together alone in the same direction.

5. That the molecules of the imponderable matters light and heat are *vastly less*, and move with inconceivably *greater velocity*, than the molecules of any ponderable substance ; and that their substance is of such a nature as to allow them to become virtually more or less oblate, in proportion to the intensity of their motion.

6. That the molecules of imponderable matters, from their extreme minuteness, are capable of *pervading and operating within* ponderable molecules ; that the intensity of the motion of the molecules of imponderable matters influences the motions of the molecules of ponderable bodies ; and that the molecules of imponderable matters thus appear in the character of *agents*.

Page 74.— *Diffusion of gaseous bodies.* There are three distinct modifications or cases of gaseous diffusion, of which, the following brief remarks will convey some notion to the reader.

First. The diffusion of *two different gaseous bodies* of the *same*, or of *different* specific gravities.

Secondly. The diffusion of *a gaseous body* and *a vapour*.

Thirdly. The diffusion of *two portions* of the *same* gaseous body, or vapour, having *different temperatures*, and consequently, *different specific gravities*.

First. Of these modifications or cases of gaseous diffusion, the diffusion of two different gaseous bodies is the best known, and is, we believe, the only case of diffusion that has been experimentally investigated. By way of illustrating the phenomena, let us suppose a flexible air-tight bag to be furnished with a stop-cock, and to be filled with a gaseous body having, under the same temperature and pressure, precisely the *same specific gravity* as atmospheric air. Let us now suppose the stop-cock to be opened. Immediately the gaseous body in the bag, and the atmospheric air, will begin to intermingle with equal velocities ;

so that for each molecule of the gaseous body passing outwards, a molecule of atmospheric air will pass inwards; and consequently under every circumstance of admixture, the bulk of gaseous matter existing in the bag will be the same. This is the simplest case of the diffusion of gaseous bodies, and will serve to give the reader some notion of that most remarkable phenomenon in its elementary form. Philosophers, as we have said, explain the phenomena of diffusion, by supposing that the molecules of any gas are *self-repulsive* or repel each other, in preference to the molecules of all other gases; so that each gas is, as it were, a vacuum to the other: and the mode in which this self-repulsive power may be imagined to operate, will be stated in a subsequent note. The case to be next considered is that in which the air-tight bag is supposed to contain some gaseous body, having a specific gravity *different* from atmospheric air; as for instance, hydrogen gas. In this case, on opening the stop-cock, the hydrogen in the bag, and the exterior atmospheric air will begin to commingle with a force and velocity proportional to the quantities of the gases diffused; and which quantities will be found to vary inversely as the square roots of the specific gravities of hydrogen gas and atmospheric air; that is to say, the volume of atmospheric air diffused inwards being supposed to be equal to 1, the volume of hydrogen diffused outwards will be equal to 3·8 nearly. These phenomena show that the velocity, or rate of diffusion of gaseous bodies has reference to their specific gravities.

Secondly. The law of diffusion of a gaseous body and a vapour is probably different from the law of the diffusion of two gaseous bodies; because the self-repulsive properties of the molecules of vapours and of gases are probably different. On the supposition that the velocities of diffusion between a gaseous body and a vapour, have reference to the specific gravities of the gaseous body and vapour, as is the case between two gaseous bodies; the velocity of diffusion of a vapour, (the vapour of water for instance), through air, must increase as the temperature diminishes; that is to say, as the specific gravity of

P.   O O

the vapour diminishes. Hence, in high latitudes, the velocity of diffusion must be extreme; and this extreme velocity may be supposed to cause the total quantity of water evaporated in a given time in such latitudes to be considerable, and thus to amply compensate for the small quantity held in solution.

Thirdly. The diffusion of two portions of the same gaseous body or vapour having different temperatures and consequently different specific gravities; like the diffusion of a gaseous body and a vapour, has not we believe been the subject of experiment. But on the supposition that the self-repulsive properties of the molecules of a gas or vapour at different temperatures are different; and that the velocities of diffusion between two such portions of gaseous bodies or vapours have reference to their specific gravities, we may imagine that there is a constant tendency to diffusion between portions of air of different temperatures; as for instance between the warmer and lighter air of the equator, and the colder and heavier air of the poles; between the colder and lighter vapour of the poles, and the warmer and heavier vapour of the equator; or between the colder and lighter vapour of the upper regions of the atmosphere, and the warmer and heavier vapour at the surface of the earth. Nor is it improbable that many natural phenomena which appear to us at present inexplicable, depend upon these tendencies. The diffusive powers of elastic fluids are at present very little understood, or appreciated: they constitute however one of the most interesting and important subjects in physics, and would amply repay whoever would take the trouble to investigate them.

Page 91, *note.*—Light and heat, and indeed all fluids have been stated to possess two kinds of repulsive power: that repulsive power which is common to them as fluids, and which may be imagined to result from the aggregate motions of all the molecules of which they are composed; and that self-repulsive power which we have supposed to depend on the mutual action of the contiguous or alternate molecules of fluids, when these molecules are in the relative positions which they are naturally inclined to assume, particularly in a state of motion. The marshalling however of the individual molecules, as those of light, supposed in this note, probably does not become apparent, till they approach, or pass through, some ponderable and transparent medium : and the passage of light and heat through such a medium, may perhaps be rendered more intelligible, by the following exposition of what appears to happen with respect to gaseous bodies. The force of diffusion, on which depends the rapidity of the motion of gaseous bodies, through any permeable medium, increases as their specific gravity diminishes. Thus the force, with which the lightest of these bodies, hydrogen, struggles to escape through any porous matter, is almost incredible; according to Mr. Graham's experiments, sufficient to raise a column of water from 20 to 30 inches. This rapidity of motion seems only explicable on the supposition, that the individual molecules of the gas, in their passage through narrow canals, are guarded from external and lateral influence; and are thus enabled to assume those positions which are natural to them, and in which their mutual self-repulsion is the greatest possible. Hence, a single row of self-repulsive molecules of a gas (or other self-repulsive fluid) passing through the minute apertures of a porous vessel into a vacuum; or what is analogous, into another gas having different self-repulsive powers; may be compared to a row of bullets urged by an elastic fluid, in quick succession through a gun barrel : but with this difference, that the gaseous molecules propel each other; instead of being, like the bullets, propelled by a foreign agency. The explanation now offered of the passage of the molecules of a gas through a narrow canal, or through any porous matter, may, as we have said, be applied, not only to the passage of light and heat through various media; but also to the passage of liquids through various bodies, by the processes which have been termed *endosmose* and *exosmose*. Do these forces operate in capillary attraction? Are the molecular motions of fluids the cause of those motions, which solid particles of matter diffused through them, sometimes exhibit?

Page 224.—TABLE OF TEMPERATURES. (FROM

Isothermal Zones.	Names of Places.	Position.			Mean Temperature of the Year.
		Latitude North.	Longitude.	Height in Feet.	
Isothermal Zones from 32° to 41°.	Nain	57° 8′	61° 20′ w	0	26·42°
	*Enontekies . .	68 30	20 47 E	1356	26·96
	Hospice de St. Gothard .	46 30	8 23 E	6390	30·38
	North Cape . .	71 0	25 50 E	0	32·00
	*Uleo	65 3	25 26 E	0	35·08
	*Umeo	63 50	20 16 E	0	33·26
	*St. Petersburg .	59 56	30 19 E	0	38·84
	Drontheim . .	63 24	10 22 E	0	39·92
	Moscow . . .	55 45	37 32 E	970	40·10
	Abo	60 27	22 18 E	0	40·28
Isothermal Zones from 41° to 50°.	*Upsal	59 51	17 38 E	0	42·08
	*Stockholm . .	59 20	18 3 E	0	42·26
	Quebec . . .	46 47	71 10 w	0	41·74
	Christiana . .	59 55	10 48 E	0	42·80
	*Convent of Peyssenburg .	47 47	10 34 E	3066	42·98
	*Copenhagen . .	55 41	12 35 E	0	45·68
	*Kendal . . .	54 17	2 46 w	0	46·22
	Malouine Islands	51 25	59 59 w	0	46·94
	*Prague . . .	50 5	14 24 E	0	49·46
	Gottingen . .	51 32	9 53 E	456	46·94
	*Zurich . . .	47 22	8 32 E	1350	47·84
	*Edinburgh . .	55 57	3 10 w	0	47·84
	Warsaw . . .	52 14	21 2 E	0	48·56
	*Coire	46 50	9 30 E	1876	48·92
	Dublin . . .	53 21	6 19 w	0	49·10
	Berne	46 5	7 26 E	1650	49·28
	*Geneva . . .	46 12	6 8 E	1080	49·28
	*Manheim . .	49 29	8 28 E	432	50·18
	Vienna . . .	48 12	16 22 E	420	50·54

(*) At the Places thus distinguished, the Temperatures

THE ENCYCLOPŒDIA METROPOLITANA.—ARTICLE METEOROLOGY).

Distribution of Heat in the different Seasons.				Maximum and Minimum.	
Mean Temp. of Winter.	Mean Temp. of Spring.	Mean Temp. of Summer.	Mean Temp. of Autumn.	Mean Temp. of Warmest Month.	Mean Temp. of Coldest Month.
−0·60°	23·90°	48·38°	33·44°	51·80°	−11·28°
+0·68	24·98	54·86	27·32	59·54	−0·58
18·32	26·42	44·96	31·82	46·22	+15·08
23·72	29·66	43·34	32·08	46·58	22·10
11·84	27·14	57·74	35·96	61·52	7·70
12·92	33·80	54·86	33·44	62·60	11·48
17·06	38·12	62·06	38·66	65·66	8·60
23·72	35·24	61·24	40·10	64·94	19·58
10·78	44·06	67·10	38·30	70·52	6·08
20·84	38·30	61·88	40·64	———	———
24·98	39·38	60·26	42·80.	62·42	22·46
25·52	38·30	61·88	43·16	64·04	22·82
14·18	38·84	68·00	46·04	73·40	13·81
28·78	39·02	62·60	41·18	56·74	28·41
28·58	42·08	58·46	42·98	59·36	30·20
30·74	41·18	62·60	48·38	65·66	27·14
30·86	45·14	56·84	46·22	58·10	34·88
39·56	46·58	53·06	48·46	55·76	37·40
31·46	47·66	68·90	50·18	———	———
30·38	44·24	64·76	48·74	66·38	29·66
29·66	48·20	64·04	48·92	65·66	26·78
38·66	46·40	58·28	48·56	59·36	38·30
28·76	47·48	69·08	49·46	70·34	27·14
32·36	50·00	63·32	50·36	64·58	29·48
39·20	47·30	59·54	50·00	61·16	35·42
32·00	48·92	66·56	49·82	67·28	30·56
34·70	47·66	64·94	50·00	66·56	34·16
38·80	49·64	67·10	49·82	68·72	33·44
32·72	51·26	69·26	50·54	70·52	26·60

given are the result of at least 8000 observations.

TABLE OF TEMPERATURES

Isothermal Zones.	Names of Places.	Position.			Mean Temperature of the Year.
		Latitude.	Longitude.	Height in Feet.	
Isothermal Zones from 50° to 59°.	*Clermont . .	45° 46′	3° 5′ E	1260	50·00°
	*Buda . . .	47 29	19 1 E	494	51·08
	Cambridge, (U. S.)	42 25	71 3 w	0	50·36
	*Paris	48 50	2 20 E	222	51·08
	*London . . .	51 30	0 5 w	0	50·36
	Dunkirk . . .	51 2	2 22 E	0	50·54
	Amsterdam . .	52 22	4 50 E	0	51·62
	Brussels . . .	50 50	4 22 E	0	51·80
	*Franeker . .	52 36	6 22 E	0	51·80
	Philadelphia . .	39 56	75 16 w	0	53·42
	New York . .	40 40	73 58 w	0	53·78
	*Cincinnati . .	39 6	82 40 w	510	53·78
	St. Malo . . .	48 39	2 1 w	0	54·14
	- Nantes . . .	47 13	1 32 w	0	54·68
	Pekin	39 54	116 27 E	0	54·86
	*Milan . . .	45 28	9 11 E	390	55·76
	Bordeaux . .	44 50	0 34 w	0	56·48
Isothermal Zones from 59° to 68°.	Marseilles . .	43 17	5 22 E	0	59·00
	Montpellier . .	43 36	3 52 E	0	59·36
	*Rome . . .	41 53	12 27 E	0	60·44
	Toulon . . .	43 7	5 50 E	0	62·06
	Nangasacki . .	32 45	129 55 E	0	60·80
	*Natchez . . .	31 28	90 30 w	180	64·76
Isothermal Zones from 68° to 77°.	*Funchal . . .	32 37	16 56 w	0	68·54
	Algiers . . .	36 48	3 1 E	0	69·98
Isothermal Zones above 77°.	*Cairo . . .	30 2	31 18 E	0	72·32
	*Vera Cruz . .	19 11	96 1 w	0	77·72
	*Havannah . .	23 10	82 13 w	0	78·08
	*Cumana . . .	10 27	65 15 w	0	81·86

(*) At the Places thus distinguished, the Temperatures

(CONTINUED).

Distribution of Heat in Different Seasons.				Maximum and Minimum.	
Mean Temp. of Winter.	Mean Temp. of Spring.	Mean Temp. of Summer.	Mean Temp. of Autumn.	Mean Temp. of *Warmest* Month.	Mean Temp. of *Coldest* Month.
34·52°	50·54°	64·40°	51·26°	66·20°	28·04°
33·98	51·08	70·52	52·34	71·60	27·78
33·98	47·66	70·70	49·82	72·86	29·84
38·66	49·28	64·58	51·44	65·30	36·14
39·56	48·56	63·14	50·18	64·40	37·76
38·48	48·56	64·04	50·90	64·76	37·75
36·86	51·62	65.84	51·62	66·92	35·42
36·68	53·24	66·20	51·08	67·28	35·60
36·68	51·08	67·28	54·32	69·08	32·90
32·18	51·44	73·94	56·48	77·00	32·72
29·84	51·26	79·16	54·50	80·78	25·34
32·90	54·14	72·86	54·86	74·30	30·20
42·26	52·16	66·02	55·76	66·92	41·74
40·46	54·50	68·54	55·58	70·52	39·02
26·42	56·30	82·58	54·32	84·38	24·62
36·32	56·12	73·04	56·84	74·66	36·14
42·08	56·48	70·88	56·30	73·04	41·00
45·50	57·56	72·50	60·08	74·66	44·42
44·06	56·66	75·74	60·98	78·08	42·08
45·86	57·74	75·20	62·78	77·00	42·26
48·38	60·80	75·02	64·40	77·00	46·40
39·38	57·56	82·94	64·22	86·90	37·40
48·56	65·48	79·16	66·02	79·70	46·94
64·40	65·84	72·50	72·32	75·56	64·04
61·52	65·66	80·24	72·50	82·76	60·08
58·46	73·58	85·10	71·42	85·82	56·12
71·96	77·90	81·50	78·62	81·86	71·06
71·24	78·98	83·30	78·98	83·84	69·98
80·24	83·66	82·04	80·24	84·38	79·16

given are the result of at least 8000 observations.

Page 224.—Explanation of the Map.

The accompanying Map, on which are traced the different isothermal lines, is taken from the article Meteorology in the Encyclopædia Metropolitana. The general character of these isothermal lines, which are founded, for the most part, on the data in the preceding tables, is, that in Europe they are convex, and in Asia and America concave, towards the Pole; and that they gradually become less and less convex and concave, as they approach the Equator.

Page 274.—On opening the fold of the Map, we see, on the left hand side, a section of a mountain at the Equator, extending to the limits of perpetual snow in that region. The right hand side is supposed to represent the Earth's surface from the Equator to beyond the Arctic circle. At the parallel of latitude of Ben Nevis, a single mountain is sketched; with the view of indicating, that though the top of Ben Nevis rises considerably above the limit of perpetual snow, according to theoretical laws deduced from the height of perpetual snow at the Equator; yet that in reality, the top of Ben Nevis, the highest mountain in Great Britain, is under the limit of perpetual snow.

Page 309.—*Of the effects of foreign bodies in the atmosphere.*—Many years ago, particular circumstances led us to form the opinion, that a combination of water and oxygen is a frequent, if not a constant, ingredient in the atmosphere. This ingredient, which we suppose to be a vapour, and analogous to (we do not say identical with) the deutoxide of hydrogen, may be imagined to act as a foreign body, and thus to be the cause of numerous atmospheric phenomena, which at present are very little understood. Among such phenomena are those of evaporation considered in the text.

When treating of the composition of atmospheric air, we observed that the best analyses almost invariably indicated a slight excess of oxygen above the amount of 20 per cent., which there ought to be in the atmosphere, if its composition were, as there can be little doubt that it is, determined by the laws of chemical proportions. Now this excess of oxygen in the atmosphere, we have every reason to think, becomes periodically associated in some way with the vapour which is also in the atmosphere ; and thus not only modifies the properties of the vapour, but at the same time materially influences the rate of evaporation from the earth's surface. This excess of oxygen may operate in the following manner. The vapour in union with oxygen (deutoxide of hydrogen ?) ceases, of course, to act as vapour ; hence in air saturated with vapour, and as moist as possible, if a portion of the vapour were suddenly to combine with oxygen, the air would as suddenly *appear* to become dry, though in reality it contained the same quantity of water in solution as before. Moreover the rate of evaporation would be increased by such a combination of vapour and oxygen ; for its effects, whatever these might be, would be superadded to the ordinary eff'cts of atmospheric air in producing evaporation, and would thus more or less increase the quantity of water converted into vapour.

Oxygen in this state of combination with vap ur seems to be particularly grateful, if not necessary to animal life. The air in which it abounds is dry, bracing, and exhilarating ; while the

P. P P

predominance of moisture, from its occasional and sudden abstraction, induces the opposite feeling of dulness and listlessness.
It is probable that some soils and situations are more favourable
than others to its existence, and that places are more or less
healthy according as it is present or absent.

The oxygen and vapour in this combination are so feebly
associated, that they appear to be separated by the slightest
cause.   Hence the results of every *common* analysis and examination of air, are the same nearly as if such a state of combination did not exist.   We may mention, however, as corroborative
of our opinion, the bleaching qualities of dew, and of the air
itself; as also the large proportion of oxygen sometimes contained
in snow water and in rain water; attention being at the same
time directed to the well known bleaching qualities of the deutoxide of hydrogen.

Much more might be said on this curious subject, especially
regarding its relation to the *electricity* of the atmosphere.   But
we desist for the present; as the difficulties attending an investigation of the atmosphere, and more than all, the total want of
opportunity, have rendered us unable satisfactorily to verify the
opinion we have advanced; which opinion therefore, is to be
understood as conjectural only; and as stated with the view of
drawing the attention of those more fortunately situated, to so
important an inquiry.

Page 374.—The table follows, illustrating the distribution of
plants over the globe, to which we have referred in the text.
It has been copied immediately from Lindley's Introduction to
Botany.

GROUPS.	Equatorial Zone, Lat. 0°—10°.	Temperate Zone, Lat. 45°—52°.	Frozen Zone, Lat. 67°—70°.
Agamæ (*Ferns, Lichens, Mosses, Fungi*)	{ Plains $\frac{1}{15}$; Mountains $\frac{1}{5}$ }		$\frac{1}{1}$
Ferns alone	{ Countries nearly flat $\frac{1}{20}$; Countries very mountainous $\frac{1}{3}$ to $\frac{1}{8}$ }	$\frac{1}{70}$	$\frac{1}{25}$
Monocotyledones	{ Old Continent $\frac{1}{5}$; New Continent $\frac{1}{6}$ }	$\frac{1}{4}$	$\frac{1}{3}$
Glumaceæ (*Junceæ, Cyperaceæ, Gramineæ*)		$\frac{1}{8}$	
Junceæ alone	$\frac{1}{11}$	$\frac{1}{90}$	$\frac{1}{4}$; $\frac{1}{25}$
Cyperaceæ alone	{ Old Continent $\frac{1}{400}$; New Continent $\frac{1}{22}$ }	$\frac{1}{20}$	$\frac{1}{9}$
Gramineæ alone	$\frac{1}{50}$	$\frac{1}{12}$	$\frac{1}{10}$
Compositæ	{ Old Continent $\frac{1}{14}$; New Continent $\frac{1}{8}$ }	{ Old Continent $\frac{1}{8}$; New Continent $\frac{1}{10}$ }	$\frac{1}{13}$
Leguminosæ	{ Old Continent $\frac{1}{12}$; New Continent $\frac{1}{10}$ }	$\frac{1}{18}$	$\frac{1}{35}$
Rubiaceæ	{ Old Continent $\frac{1}{14}$; New Continent $\frac{1}{21}$ }	$\frac{1}{60}$	$\frac{1}{80}$
Euphorbiaceæ	$\frac{1}{32}$	$\frac{1}{90}$	$\frac{1}{500}$
Labiatæ	$\frac{1}{40}$	{ America $\frac{1}{40}$; Europe $\frac{1}{25}$ }	$\frac{1}{70}$
Malvaceæ	$\frac{1}{35}$	{ Europe $\frac{1}{200}$; America $\frac{1}{100}$ }	$\frac{0}{0}$
Ericeæ	$\frac{1}{130}$	{ Europe $\frac{1}{100}$; America $\frac{1}{36}$ }	$\frac{1}{25}$
Amentaceæ	$\frac{1}{800}$	{ Europe $\frac{1}{45}$; America $\frac{1}{25}$ }	$\frac{1}{10}$
Umbelliferæ	$\frac{1}{500}$	{ Europe $\frac{1}{40}$; America $\frac{1}{18}$ }	$\frac{1}{60}$
Cruciferæ	$\frac{1}{800}$	{ Europe $\frac{1}{60}$; America $\frac{1}{18}$ }	$\frac{1}{14}$

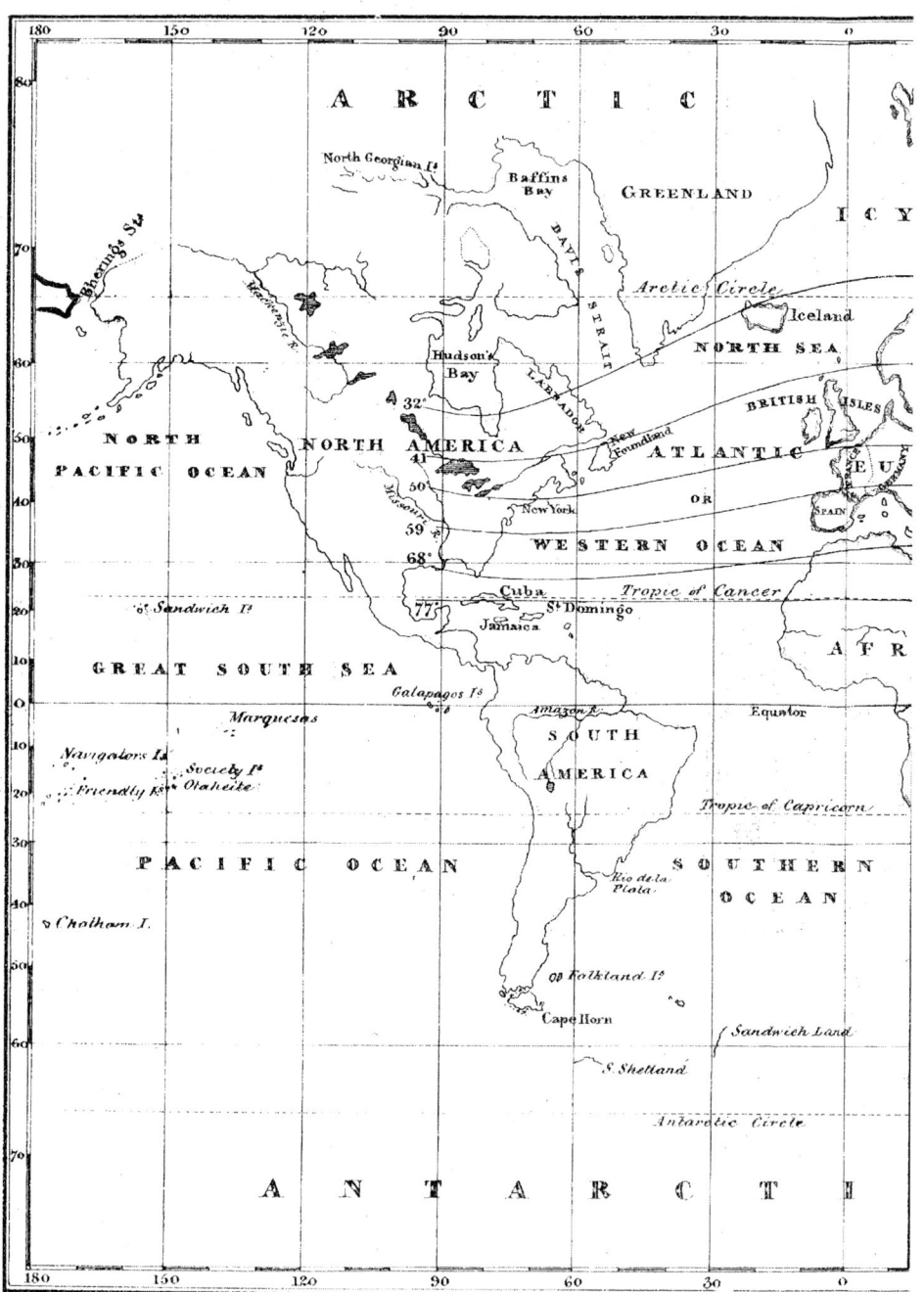

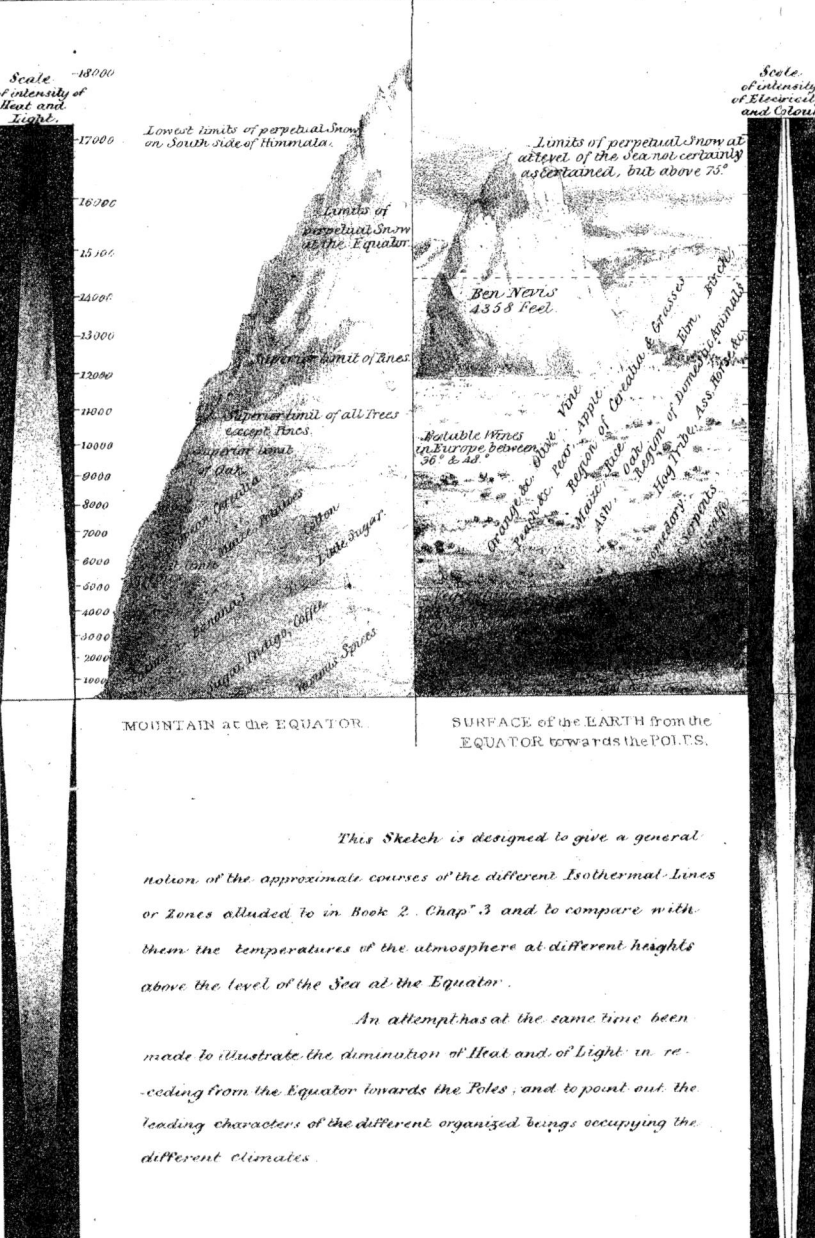

Scale of intensity of Heat and Light.

−18000
−17000
−16000
−15000
−14000
−13000
−12000
−11000
−10000
−9000
−8000
−7000
−6000
−5000
−4000
−3000
−2000
−1000

Lowest limits of perpetual Snow on South side of Himmala.

Limits of perpetual Snow in the Equator.

Superior limit of Pines.

Superior limit of all Trees except Pines.

Superior limit of Oak.

MOUNTAIN at the EQUATOR.

Limits of perpetual Snow at a level of the Sea not certainly ascertained, but above 75°.

Ben Nevis 4358 Feet.

Notable Wines in Europe between 36° & 48.

Orange &c. Olive. Vine. Apple. Region of Cereals & Grasses. Elm. Birch. Region of Domestic Animals. Hay Tribe. &c. Forests.

Maize. Rice. Oak. Ash.

SURFACE of the EARTH from the EQUATOR towards the POLES.

Scale of intensity of Electricity and Colour.

This Sketch is designed to give a general notion of the approximate courses of the different Isothermal Lines or Zones alluded to in Book 2. Chap 3 and to compare with them the temperatures of the atmosphere at different heights above the level of the Sea at the Equator.

An attempt has at the same time been made to illustrate the diminution of Heat and of Light in receding from the Equator towards the Poles, and to point out the leading characters of the different organized beings occupying the different climates.

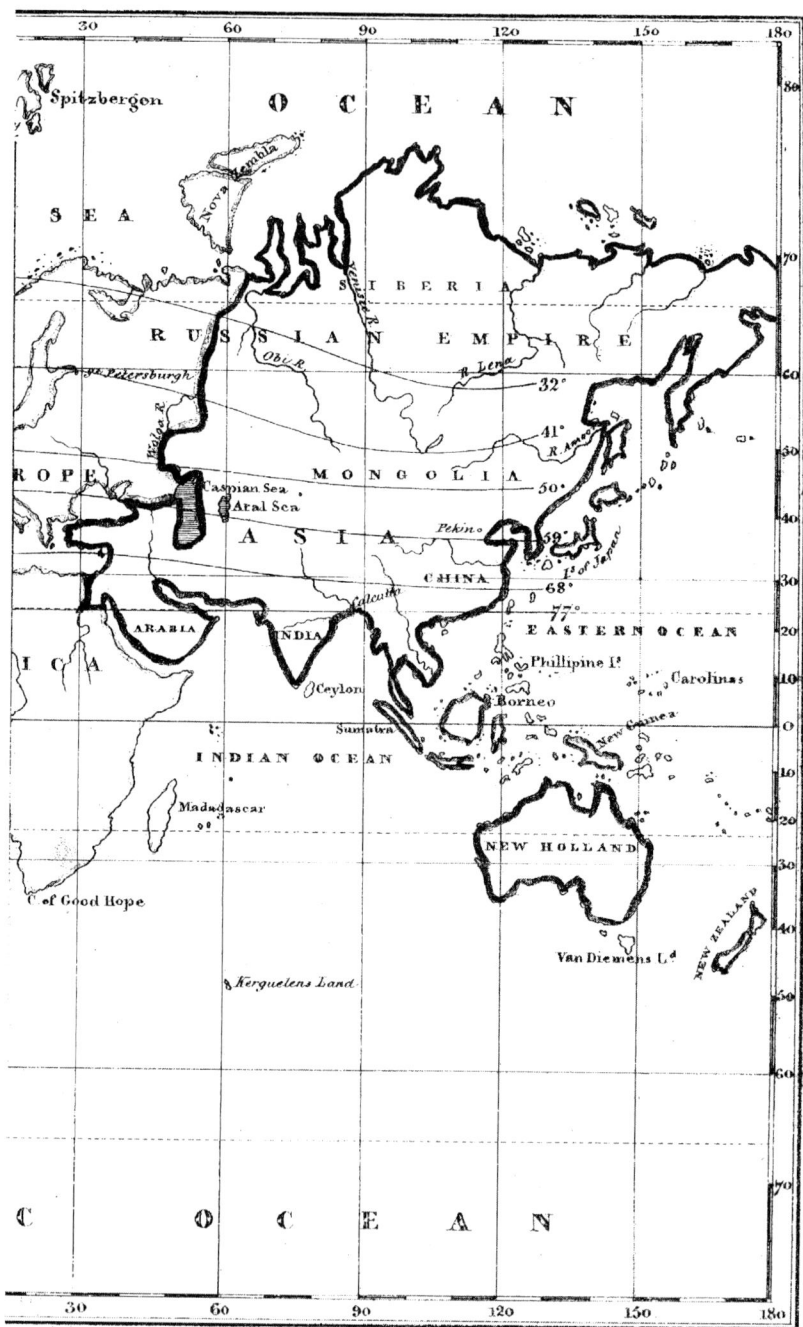